ELEMENTS
OF THE GREAT
CRYSTAL

JEAN DALTON

Author Email address required. **catrinalynne123@gmail.com**

Acknowledgment

My brother **David,** for pushing me to keep going, and he was the first to believe in me besides my friend, who helped me get it done. **Melo, Moogs, Muse** and **Cookie** helped me when i was down, they always brought me back up.

Dedication

I dedicate this book to my Daughter **Catrina**, who means the world to me, she's always been so positive and was the first to read my book, I love her so much, She is the best.

David. He is one of my favorite people and I wouldn't be so driven to write without him.

My mom **Barbara** and my dad **Edward,** and my step mom, **Mar**, for teaching me so much about life.

Debbie, my sister who got me into books in the first place, she changed my world, because books take you away from reality, and that I can never say too much about how respectable and caring **Debbie** is..

There are a few teachers that really reached me. First was **Mr Holmes,** then **Mr wolf,** then **Mr Alexander, Ms rau.** These teachers taught me so much, they treated me great and I hope they are doing great.

TABLE OF CONTENTS

About the Author

I grew up in the 80s, where everything was so good, that's got to be my favorite decade. In the city of Windsor Ontario Canada, I lived there most of my life. I went to Hugh Beaton, grade school, then high school I went to Kennedy C. I. loved that school. I grew up reading, my first book was "Comes the blind fury" , then I read Cujo, and became addicted to Stephan King's book.s, though I haven't been able to read much.

Melissa M would write her stories, and I loved them.. She also lent me so many books, I'd read them and want more. like a book a week if even that. IN high school, all I did was make comics. but one of my sisters threw them out so i stopped, i had over 500 pages of that comic. she just threw them out like they were nothing. which broke me, but now i'm writing, which in my opinion, is a lot easier, and it's just so much fun to write. I have a beautiful daughter, shes the best thing to happen to me. She makes me proud every day., and she helped so much.

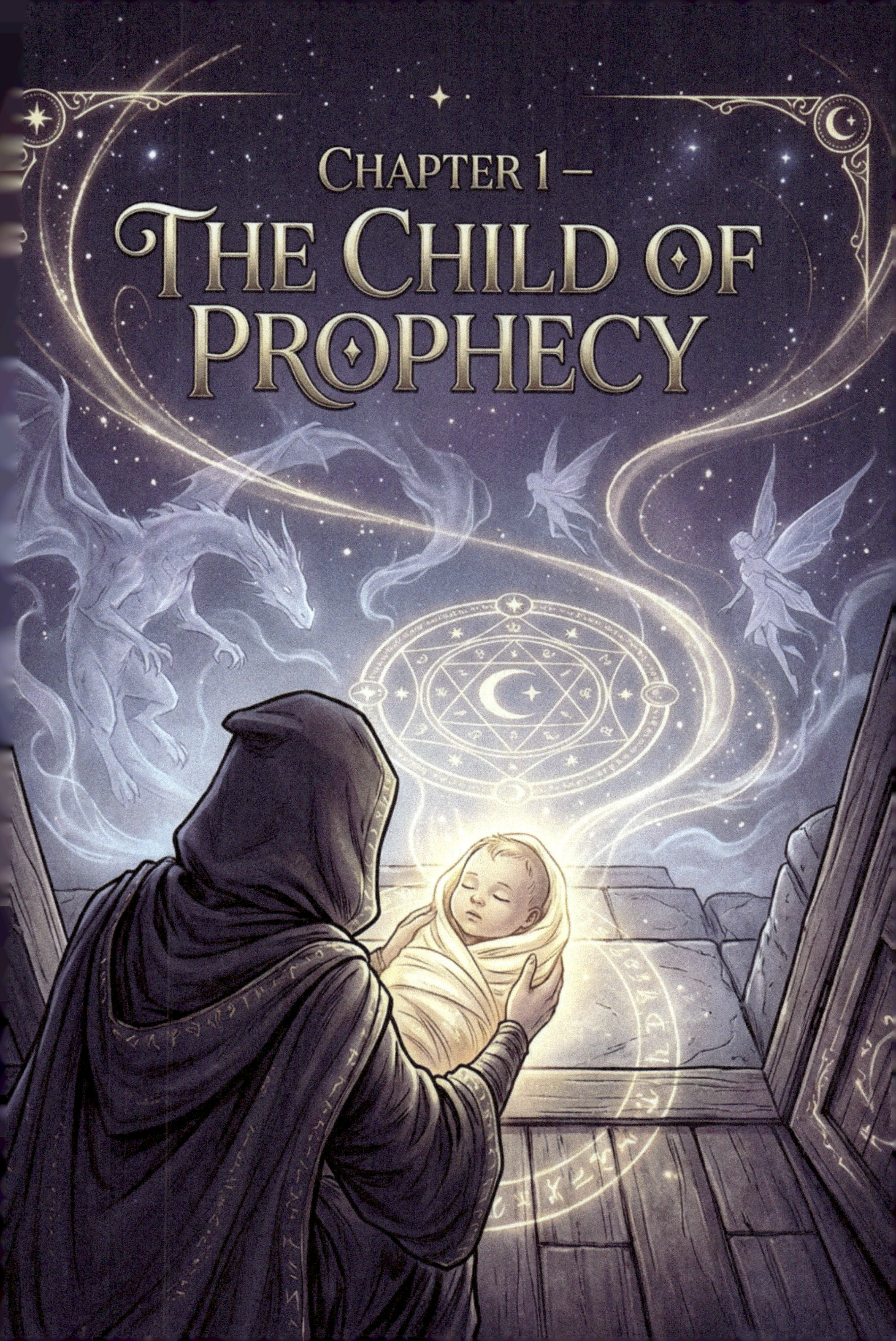
CHAPTER 1 –
THE CHILD OF PROPHECY

CHAPTER I

THE CHILD OF PROPHECY

There once was a world full of magic and magical creatures, from dragons, phoenixes, and unicorns, to faeries and elves. They all lived happily among humans. The good magic prophesied a great evil taking over. These people and creatures feared that they would be captured and their magic used for evil. They didn't want their magic abused or used to hurt the humans they loved. They decided to leave for a new dimension so that they could continue in peace. That would leave only humans to deal with the issue, because humans were strong against evil. But they felt a great sadness; they couldn't bring humans with them, for they wouldn't survive the trip. A new beginning must be forced to happen in time if the world is to be saved. But the prophecy continued, saying that when evil grew, they would leave a child behind, and it would live as a human, unaware of its own power.

But all of that could take hundreds of years. They taught magic to humans who became known as witches. There were a few elder witches who taught apprentices, so some magic would remain on the planet.

Two hundred years passed. In a faraway small town called Koabi, there lived a married couple. The husband was the greatest fighter in the land. The wife, a kind woman who designed and made clothes. They could not bear any children, which destroyed him and his beloved wife. Only a chosen few knew they were the perfect couple to take on this child from a different world who was the future of their kind and the future of Eterial. All the magic of Eterial left over two hundred years ago. This would be a tough journey for the child, but if set right, the prophecy would come true and make the world magical once again.

The hooded person left the baby wrapped in purple silk with a note that simply read, “Please help me.” And that was all; that was all they could say, so they could only hope this would set the prophecy. The hooded figure knocked on the door, then vanished.

“Who would be knocking on our door so late?” Luthor growled.

“Who knows with the world today?” Gwen laughed and threw a pillow at her grumpy husband. “Go answer it already, my big, strong husband.”

“All right, if it’s a drunk lost, I might have to get my sword!”

“No, you won’t,” Gwen giggled.

Luthor walked to the front door and opened it. He looked around, but saw no one.

“What the heck? Why would someone knock and then leave right away?”

The child gurgled and cooed at the man.

Luthor looked down and was shocked to see an infant on his doorstep.

"Honey, could you come here please, it's important!"

Gwen rushed to him, then looked and was not happy.

"What, no emergency?" Gwen said, annoyed. Then looked down after hearing the baby gurgle. "Oh my goodness!" Gwen swooped the baby into her arms. "There's a note they read and realised thisbaby was a miracle to their family. Just yesterday, they received a letter that said she was infertile. The child needed their help, and they desperately wanted a family, but they couldn't have one. No one was to know this child was not theirs; they would hide it, then act like the baby was a miracle baby, and that she didn't even know she was pregnant. If anyone dared to think differently, they would make it known; it was theirs by force if need be. But they kept to themselves, and the neighbours were pretty far away, so who's to say they didn't have a baby?

They brought the baby in and laid it on the bed. Gwen unwrapped the baby and was happy to find it was a girl.

"I have the perfect name for her, it's the name I chose if we ever had a girl, Lillith!" said Gwen.

"I like that name, it has character!" Luthor kissed Gwen on the cheek.

"Look at the marking on her forehead, it's a star shape, a six-sided star, and is a strange color, lilac? It may go away in time, probably just a birthmark, I have already fallen in love with our little Lillith," Gwen purred, so content with this new addition to their family.

“I feel the same, my dear.” He put his arm around Gwen.

Then suddenly, as if Lillith was saying she agreed, she urinated all over the silk and their nice clean bed.

“Oh dear!” Gwen snorted.

“She approves of us and her name! For she showed through urination!” Luthor laughed. He then picked up Lillith and let Gwen gather the silk and blanket off the bed.

“Your mother has a lot to do tomorrow! Make you some new clothes, my darling, first thing in the morning. I’ll buy the comfiest cloth to make some outfits for you! Right, dear?”

“Of course! And luckily, we kept the crib!”

They couldn’t part with the crib they had gotten because they had tried hard to make their own child, and it had only been a day since the letter came in from two towns over saying she was infertile because she was missing some vital things. Which broke her heart, and she had cried when they heard because the doctor had to do research, and they hoped everything would be fine, but the letter was so blunt. Gwen felt so guilty. She told her husband he could leave her and she would understand, but her husband loved her and would never leave her. They kept the bad news a secret, not even telling their parents. The only one who knew was the doctor, an old man who was about to retire. Although he was blind and not the brightest, he was an expert on birthing a child. But many would think, Oh, the old guy made a mistake, big deal! So they were glad for that.

Luthor hugged his darling new baby Lillith and cradled her softly. "Do we have milk? Or even a bottle? She looks hungry!"

"Yes and yes!" Gwen boiled the bottle, warming up the milk for her baby. Oh, it felt good to think they have a baby now!

She took the bottle out, tested it on her wrist, and it was just perfect!

"May I feed Lillith? Please," Gwen smiled with puppy-dog eyes.

"I can not say no to those eyes!" Luthor grinned and handed little Lillith to her new mommy.

"Okay, little one, we have a yummy treat for you. She placed the nipple of the bottle to Lillith's lips. They waited, and waited holding their breath, in fear she might not accept the milk and starve. Last second, Lillith moved her lips, then sucked the nipple, then gurgled happily as she started to drink her first drink with them. She finished off the bottle, Gwen put Lillith to her shoulder, and patted her back until there was a big burp.

"Hahaha! She takes after you!" Luthor laughed.

"Very funny, now get me a pillow case, I must make a diaper for Lillith, so she doesn't make another mess, one of the thick winter ones, please, for maximum protection."

"Yes, my wife!" He went into the closet and pulled out a very nice pillow case. He watched as Gwen transformed the pillowcase into a diaper.

"There, my little Lillith, you are all set for bed." After saying that, Lillith yawned. This was amazing! They took out a warm, soft towel.

"Ok, Luthor, rock your baby to sleep!" Gwen knew this was what he always wanted and was happy to grant his wish. Luthor gently took Lillith and rocked her back and forth slowly, singing some nonsense song Gwen had never heard before, so she assumed he was making it up as he went.

Lillith slowly drifted off to sleep. He placed Lillith into the crib, and the wife covered her with a soft, warm towel.

They both giggled and jumped into their now blanketless bed, sliding under the sheets.

"What a great day it is to be alive, my dear," Luthor said with a yawn.

"This secret will not ever be repeated to even Lillith, okay?" Gwen said.

"Yes, my wife, our family is now complete."

"Yes, she is a gift from the Gods."

****

A couple of weeks later, they decided to show the world their new baby. They showed her off to everyone, and everyone said she was the most beautiful baby they had ever seen. Lillith's eyes were still newborn blue, but she had some red hair appearing on her head. Neither of them had red hair, but they could make excuses. They were not worried. They walked down the street, houses scattered on either side of the street.

"Oh, she looks just like her mother!" Said the old woman who lived down the street from their house, "They never got to know the people in this town. They had moved here so the husband could open a school on fighting, since there were no wars for many years."

He used to be the leader of the knights, but the kingdom fell, and the people were happy it did. They had caused nothing but turmoil and taken most of their money for taxes. He was a royal pig, hogging all the best food for himself and his very fat court. Luthor was supposed to fight for the king, but what kind of selfish being deserved to be a king? They needed no kings or queens! The people ruled over themselves, and everyone was happy with that. It was Luthor's doing that everyone chased the royalty and snobs out of power, and they fled to the second continent.

Money was earned through jobs and used for food, clothes, and other necessities. But sometimes people would trade food for clothing or for a hard day's work. It was very peaceful and calm throughout the land.

There was no king in all the world; that's just the way it worked. There were castles all over Eterial, filled with cobwebs and dust. The second continent is called Chella. They were having problems over there, with something to do with evil, but what could they do? It wasn't spoken of much, so the people in charge thought it wasn't a big deal, that they could handle themselves.

"Luthor, did you hear?"

The father turned to look at his friend, who had moved to Koabi to help teach.

"What is it?" He then tried to show off his daughter, but his friend wasn't the brightest, but a damn good fighter!

"There is some sort of evil on Chella across the great ocean. I heard one town was destroyed, but only one town."

"What? Really? EVIL?" He huffed. "It's better to train fighters now, so if it happens to come here, we can fight it off!"

"That's exactly what I thought!" He smiled.

Luthor then pointed to Lillith.

"What's this? You had a child? Congrats mate! I never even noticed Gwen being pregnant!"

"She was a surprise to all of us, but I'm glad you didn't notice the lump, gotta keep my girlish figure." Gwen laughed. Then kissed Lillith on her forehead.

"Is it a she or a he, and what's its name?"

"Her name is Lillith," Luthor said proudly.

"What a beautiful baby, and quite a strong name! I wish you both the best!" He then ran off, probably to gossip more on this so-called evil.

The couple paraded little Lillith around Koabi.

"That baby has an ugly mark on its forehead, is it dumb?" A shocked mother heard her four-year-old say.

"Oh my goodness, excuse my son, he is not aware of birthmarks!!" She bowed her head and encouraged her child to follow suit. "You have earned yourself a spanking when you get home!" She mumbled into his ear. He began to cry.

"No need to punish the boy," Gwen said kindly. "He is the first to comment on it. I mean, he called her dumb, but it's something new to him, so a child his age might think she's different or that something is wrong with her. So please don't punish him for being open and blunt," said Gwen, smiling at the young boy.

The woman stood up straight and smiled. "Thank you for your forgiveness. He always notices the littlest details in all the different things. You hear that, son, you are smart and honest."

The boy stopped crying and started to smile, wiping the tears from his eyes.

The happy couple continued along their way when an old lady stopped them. They could tell she was the Elder witch.

"May I please touch the child's forehead?" She asked sweetly.

Luthor shrugged at his wife, Gwen.

“O-ok,” Gwen said, a little nervous.

She gently placed her hand on Lillith’s forehead, and a wide smile spread across her face as she jumped a little. Then, she cackled under her breath. “Thank the Gods.....I knew it was starting,” she removed her hand.

“You just made an old woman’s day! “may I give you something for putting a smile on my face.”

“Sure,” Luthor smiled, but was confused.

She handed a small string of purple beads. “Hang this over your door to your house, it will keep evil away; they will not be able to enter.”

Luthor and Gwen looked at each other. Then smiled, because they believed in charms.

“Thank you,” Gwen said, and Luthor took the beads. She was still smiling as she backed away, and she was laughing and cheering them on. They laughed and said thank you again before continuing on their way. They stopped at a pub, Luthor ordered a steak, and Gwen ordered a few slices of ham. Before Gwen started eating, Lillith began to fuss.

“I know my sweet, I know you are hungry too!” She took out a bottle of milk from her bag and fed her.

After finishing their supper, they went home, changed her into a new diaper, cleaned her up, and then put a fresh one on. They tucked her into bed with a real baby blanket.

A lot had changed; they had everything they needed for Lillith. Then Luthor hung the beads above the door with a nail and yawned. It was a long day. Gwen changed and got into bed. Luthor followed and kissed her on the cheek. Good night, they both said. Luthor was already snoring.

****

Lillith was now a big girl, at the age of five. She had already started practising fighting with the wooden sword her dad had made her. Her eyes were now the same color as her birthmark, a warm lilac. She accompanied her father to his lessons on fighting, although she couldn't participate. However, she enjoyed carrying her sword and would try to copy her father as he taught thirty young men in each class. She tried so hard to copy her father, but she just couldn't get it right, and you wouldn't know that by watching her. She felt she was doing exactly what he was doing. She was her daddy's little munchkin. When he gave her lessons, she would do them perfectly in a day or two. She never stopped practising. She often wore her hair up in pigtails. With her fiery red hair, she went for a walk and came upon a group of seven-year-olds. They laughed at her for carrying a wooden sword around.

"What are you, a baby? And look at that horrible birthmark, it's not even red like normal ones! You are ugly!"

Tears welled up in her eyes, she then wiped them and her sadness turned to anger, which turned to rage, she walked right up to the boy and smacked him hard on the butt with her wooden sword, then ran home which was on the same street they were on, but there was a big gap where houses stopped and her home was. She lived right next tothe forest, but you could still see her house from his. The boy started crying so loud, his mother came running.

“What happened?”

“That girl just beat my butt with her stupid wooden sword.”

“Well, let’s go have a nice little talk with that brat’s parents!” She petted her precious son’s head. He was still crying because it hurt. They saw where she lived by watching her run home, not far from his own house. She stomped right up to the house and banged on the door. Gwen and Luthor answered the door.

“May we be of help to you?”

“Oh, you bet you can be of some help, your child just beat my boy with her sword on the butt!”

“What?” Luthor turned into his home and called his daughter. “Lillith, can you come here right now?”

Lillith came without her sword, which was strange for her. She secretly hid her sword so her daddy and mommy couldn’t take it away. She was wiping her eyes like she was crying, which she was. But she was trying to hide it.

“Why did you hit this boy with your sword?” Luthor asked her.

“I didn’t hit him hard!” Lillith pouted.

“Why, though, you had to have a reason, my daughter doesn’t hurt people, especially with a weapon for no reason.”

The boy gulped and became nervous.

“He said I was a baby for carrying my sword, then made fun of my birthmark and called me UGLY!!!” Lillith ran away crying hard. Gwen followed Lillith.

“Oh, I see,” Luthor glared at the boy, and the boy now recognized the father as the toughest man in all of the land, and so did the mom.

“You did that to her, you little brat!! You think that sword hurt your butt? Wait till you tell your father that you insulted his teacher’s daughter!!!” The boy began crying. He knew he had screwed up, and he wanted to join that school one day when he was older; he felt lower than dirt.

“Sir?” The boy put aside his mom and didn’t care about his spanking.

“Yes, young man.”

“Can you tell your little girl, I was trying to impress the cool kids, I think she’s beautiful, I like her birthmark, wish I had one like it, but maybe a circle? And I think her sword, even if it’s wood, is a mighty weapon." He rubbed his butt.

Lillith heard him say these things, and she ran to the boy and hugged him. "You are now my friend!" She announced. "My name is Lillith.

What is yours?" Lillith didn't have any friends, but she wanted to be friends with this kind boy.

"Andrew." They both started laughing, then all three parents started laughing, like some group of idiots. The mom said sorry, patted Lillith, and told her she was so pretty and smart to stand up to bullying. The boy kissed Lillith on the cheek and ran home. Meanwhile, the mom walked to her home and didn't mention this to Andrew's dad.

The two hung out every day, except when she was being trained. Being as strong as Daddy was her goal. But she always wanted to have fun with Andrew.

****

Lillith was now ten and was getting stronger and stronger, a better fighter with and without a sword. Her father took on Andrew to train as well; he was really good with a sword, and was quite the fighter too. He was nothing compared to Lillith; she already surpassed his best fighters, which made no sense to him, but boy, was he proud. She would lose against them, because she was four foot seven to their six foot something. But her form and strength were far greater than those of a ten-year-old.

Andrew wasn't half bad himself; he could hold his own against kids his age, but he had special training, so he had an advantage.

They were both very quick on their feet, but Lillith was quicker. She always seemed to do better, no matter how hard Andrew tried. It was like watching Lillith dance; her moves were so smooth. Andrew was great, but Lillith was outstanding.

“Awe, come on, you can’t always win!” Andrew whined.

“Don’t worry, I will protect you,” Lillith giggled.

“Well, you have five years on me in training....”

“Don’t feel bad, Dad said I could beat his best-trained fighters if I were as tall as they! It’s nothing against you, I just learn quickly and like dad says, I’m a natural, but dad says you are too!” Lillith explained.

“Really? He said that?”

“YUP!” Lillith smiled.

She started to turn away, and Andrew went in for an attack. But Lillith blocked it so damn fast that it knocked the air out of Andrew, and he fell backwards.

“Oh my God, are you okay?” She went to her knees, leaning over him. He caught his breath, then looked into her purple eyes, and leaned in to kiss her. She leaned in, too, and they had their first kiss. It felt nice, but they decided to become a couple, but it’s a secret, no one must know!

Yes, she’s ten, and he’s twelve, but come on, puppy dog love!

They were best friends and secretly in a relationship. They weren't acting upon it; they never kissed, instead just holding hands when people weren't looking.

They both had dull metal swords to use; they spent time sparring and practised fighting hands-on. She liked that he didn't get angry because she never let him win. He did pout, but he understood she was different. He didn't fully understand what that meant yet, but would probably figure it out when he got older.

Today was Sunday, which meant no fighting allowed, because every warrior in training needs a day to rest, says Lillith's dad.

"Let's go into the forest! We can catch frogs or I dunno, discover treasure?" Andrew suggested.

"Sounds like fun! I hope I catch a lizard! I heard if you scare them, their tail falls off, I want that, I want it bad!" She spoke with pure wonder.

"Sounds like a plan, an operation to find a lizard and scare them until their tail falls off!" They laughed and both walked towards the big old trees. This was going to be fun. They looked everywhere, on trees, in holes, under rocks. This mission was much harder than they thought. The forest was huge and perfect for kids to play in, with low branches, lots of beauty, and a safe environment for them. The forest was incredibly old, with huge trees that kids could climb and have fun on. It was a blast to play there.

Lillith looked at the position of the sun. "Break time, they climbed onto a branch in a big tree, it must have been hundreds of years old, they climbed up above the ground, ten feet in the air, it was a perfect place to eat and drink. “It’s lunch, and mom made us roast beef sandwiches!!!!! With a large bottle of milk, one for you and one for me! Isn’t that awesome of Mom! I love milk. She handed him the wax paper with his sandwich in it, then grabbed hers. She dug in, eating like she had never eaten before.

“Yikes, you don’t eat lady like,” he laughed.

“You are just noticing this now?” Then stuck out her tongue at him.

“I never paid attention, but I honestly don’t care how you eat, I like you for you....”

Lillith blushed. “Well, thanks, I don’t care about anything bad you do, as in manners or hygiene.”

“I shower often!” He shouted at her, then took another bite of his sandwich.

“I’m just kidding, I just didn’t know what to say.” She took a huge bite, monster chewing, until she swallowed.

“You are lucky you are pretty, or I’d have to leave you, for fear of being eaten.” Andrew laughed.

Lillith giggled. “I just don’t care, Dad says as long as you chew it enough and swallow fine, it’s eating!”

“Yea, please don’t choke, okay?” He looked serious.

"Fine, I will take small bites." Lillith took a small, regular bite of her sandwich.

They finished their sandwiches, and Lillith looked in the bag, shocked.

"Mom packed a piece of cheese with our lunch, and some beef jerky!" She squealed.

"Oh my, I'm gonna have a full tummy for a while! GIMMY GIMMY GIMMY GIMMY!" Andrew ordered playfully.

Lillith handed him a chunk of cheese and beef jerky, then grabbed hers. She took small bites just to prove she could eat normally. She finished, then drank the milk, almost in one gulp, then burped loudly.

"Wow, let me do the loud burping!" Andrew drank the whole bottle in one gulp, then burped so loud it echoed.

"Good one!" Lillith laughed, then saw something move. She turned her head and saw an orange lizard!

"What do you see, Lillith?" Andrew wondered.

"I found a lizard, I gotta catch it, don't move!" She slowly faced the lizard, then got up slowly, then pounced and fell to the ground.

"Lillith! Are you okay?" Andrew yelled. They were ten feet off the ground on a branch.

"Yea, I CAUGHT HIM!" She was so proud of herself.

"How are you okay? You should have a broken arm or a dislocated shoulder from that fall!"

"Nah, I'm fine, musta have landed softly! Come, we must finish the mission!" Lillith demanded.

"Hmmm," he realised she had a very strong body. Andrew climbed down and looked at the lizard. "How do we scare it?"

Lillith screamed at it; it nearly died, and its tail fell off, squirming on the ground like a worm.

"Poor guy, got scared by the evil Lillith!" Andrew let the scared lizard go, then looked at the squirming tail.

"Don't worry, they grow back," Lillith said matter-of-factly. "I wonder why it squirms so much. What is the point?" She was getting grossed out staring at it.

"Oh, it does that, so the thing that scared it will chase after that, and let the lizard escape with its life," Andrew explained.

"Our mission is complete! What next?" Lillith wondered.

"I think we should head back into town."

"Sounds good, we can hide and practise hand-held fighting, maybe I will let you win?" Lillith poked Andrew. "Come on!"

They both walked together, and then they grabbed each other's hands.

****

Lillith was now fifteen; she was the bravest, toughest, quickest, and shortest warrior. Lillith was only five feet, well, four feet eleven and a half, but she says she's five feet. She was told she would be this height for the rest of her life. That drove her nuts! She wanted to be tall like her dad, or even her mom, who was five feet seven! She was annoyed because Andrew was six feet, but then the doctor said boys keep growing until sometimes their 20s. He shouldn't have told her that. Poor Andrew.... Lillith left the doctors.

"Andrew! Get ready to have your butt handed to you!" Lillith went home and grabbed a larger, heavier sword, but it wasn't sharp.

Andrew just walked to Lillith's house and sat down. He was now seventeen.

"Let's just talk this out, because I am in no mood to get my butt kicked today. What happened?"

Lillith dropped the sword and walked out to meet Andrew.

"I'm super sad..." She began to cry and covered her face with her hands.

"Hey hey hey, what happened?" Andrew asked worriedly.

"The doctor told me......I will never grow any taller....." She sobbed.

"So? What does that matter?"

"I want to be the best, and I can't, because I'm short...." She wiped tears out of her eyes and tried to calm down.

"You already are! Your dad can barely beat you right now; there is not one warrior that can beat you, besides your dad."He looked into her lilac eyes, with his deep ocean blue eyes, and smiled, then put his forehead to hers. "In a few years, you will beat your dad, and....maybe in a few years......we can get married?" He took out a ring, golden with a deep blue sapphire in it.

Lillith gasped. "Oh my God, I love it!"

"I love you." He said with a big smile on his handsome face. "Do you love me?" Still smiling.

She blushed and smiled.

"I do love you, my Andrew. And yes, I will marry you." She grabbed his head and kissed him deeply and softly. They both sat on the ground and just looked into each other's eyes. Andrew knew there was something very special about her. Over the years, her falling and not getting hurt is just not normal. There were other things, like how she could catch anyone silently attacking her. Each time she blocked them, it was like she moved so damn fast that she had to be special. He leaned on the house and pulled her into a hug, and they cuddled. Lillith's dad came home and stared at the two. "What?"

The two giggled. Lillith showed her father her engagement ring.

“WHAT!” He was happy but sad at the same time. “You guys have to wait, you can be engaged for years!”

“That’s the plan, Dad,” Lillith smiled.

“Oh, your mom is going to freak out, in a good way.”

****

Lillith dodged and then struck her dad right at the heart.

“I WON! I BEAT YOU, DAD!!! FINALLY!” Lillith laughed and jumped up and down happily.

“You know you are the only one who can do that, you should feel special! I wasn’t even going easy on you!” Luthor said proudly, and out of breath.

“I know, Dad, I’m gonna go eat dinner, I’m starving!”

“I bet.” He laughed. Wait for me!” Her father laughed and ran past her to beat her to the house.

Lillith giggled and chased after him. “I will catch you, Dad!” she laughed and jumped on his back.

“Just like the old days, I will carry you to whatever mom is cooking for us.”

Chapter 2 – The Awakening of Power

CHAPTER 2

THE AWAKENING OF POWER

Lillith was wandering through the forest. She was now eighteen, an adult. She felt something on her head; Her forehead was getting warm. As she walked forward, the warmth increased, then intensified. However, when she turned left, it grew colder.

"Okay, you want me to go right." She thought to herself, 'I'll return to where it's warm.' She went right, and it got warmer. She continued to follow her forehead, getting it wrong and right until she got to a hole in an old oak tree, and her forehead burned. Lillith looked in and found a single golden egg, a tiny blue bird-sized egg. She picked it up; it was cold.

"Oh no! It will die! If it hadn't already...." She ran home and set the egg close to the fire in the fireplace. She then designed a nest that she could wear around her chest, where her heart would keep it warm. She was glad she learned how to sew. She placed the egg softly into her homemade chest egg warmer. She loved this egg; it was one she had never seen before, and it had a unique design, made of gold. What bird could this egg hatch into? Ewww, what if it's a snake? Did she really want that to hatch and crawl out? Oh well, it's worth the risk.

****

Lillith decided to go to the pub for a few drinks, just to relax. She wondered where Andrew was. He was twenty now. He became the third-best fighter in the land, with her father second and herself first. She placed her hand over the egg and sent it love; she could feel its warmth, almost as if she could feel a heartbeat, but there was no doubt that was not possible. She walked to Andrew's house and knocked on the door. His mother opened the door.

"Oh, my future daughter-in-law! Do you want to come in and have some tea?"

"Oh, I'm just looking for Andrew. I wanted to go to the pub, have some snacks, and drink some beer. You know, the usual Sunday nights."

"Oh, hmm, he's not home, I actually think he is at your place talking to your dad."

"I wonder why?" Lillith pondered.

"Oh, prob something to do with fighting, or men stuff." She spat out.

"Oh, ok, well I'm gonna go get him, thank you, you were very helpful." Lillith smiled and turned around, starting to walk towards her house. She entered and sighed. "Sundays are our rest days, does Andrew not remember that?" She walked in and found her mother sewing a dress she had been ordered. "Mom, is Andrew here? Is he with Dad?"

"Oh yes, he was here, but he and your dad went to the pub. I'm surprised you weren't with them?"

"Ugh, that annoys me to no end! This is Andrew and my funday, why is dad cutting into our funday?" She wanted to punch Andrew really hard on the arm.

"Oh, hun, they were whispering, I dunno about what, but they left in a hurry"

"Okay, I'm out," Lillith said grumpily, and then she left.

Lillith slammed the pub door shut; everyone looked at her, including her father and her fiancé, with whom she was not happy at the moment.

"What the hell is going on? How dare you betray our funday, Andrew, and Dad!" She glared at them both. Andrew and her father moved to a table and told her to come over.

"I will be over when I get my pint."

Her father stood up, "No, come here."

She raised an eyebrow in confusion. Then walked over to them and sat down. "What's going on?" She questioned.

"Well, see, you know the continent across the great sea? It's in ruins, most ppl are okay, but many have died horrible deaths."

"What? How, why?"

"There's a group of evil creatures trying to take over, we are guessing, but they crossed the great sea and are now invading towns here. On our home continent, Ethridge."

"Oh my God, evil? We can't fight evil!"

We know a troop I trained went to fight them, but they all died horrible deaths. We don't know what to do, but we hope the evil dies out; we are just sitting ducks basically."

"What are the evil things?" Lillith questioned.

"We know the army is what you would call Wendigos, dead things, nonstop eating people and killing for pleasure, we know there are higher up creatures that are controlling the Wendigos, but we have no clue."

"What are Wendigos?"

"We are not sure, but they attack even if they lose an arm or leg. We think they are undead things summoned by the evil that's controlling them," said her father.

"Well, who is controlling them?" Asked Lillith

"We don't know," Andrew answered.

Her forehead burned, as if it were reacting to this topic.

"How many towns have been destroyed?" Lillith asked.

"Two so far, Donice and Figert, they have to regain their power, they do a town every couple of weeks, it would seem."

"I need to think, I'm going to go for a walk," Lillith said with this new news heavy on her shoulders and heart.

"May I join you?" Andrew asked.

"You never have to ask, unless I'm pissed off at you, which I was, but now I'm not."

"Okay, I just had to tell your dad, I was worried, I would have told you first, but I just felt your dad would know what to do, and he already knew most of it."

"It's ok, that doesn't bother me. What are we going to do? I feel I need to save people, but I'm powerless...I feel so useless...."

"We will figure out something."

Just then, something started squirming on Lillith's chest. She looked in and checked on the egg; it was moving.

"Oh my gods, my egg is hatching!!"

"A golden egg?" Andrew was astounded.

She shushed Andrew and watched it crack. Then boom, it popped open like it was too impatient to do the regular hatching. A small creature popped out of the golden egg; it had two horns on its head and spikes down its tail. Its wings were large, almost resembling demon wings. It was golden in color and had beautiful blue eyes. It was covered in golden scales and was incredibly small. Its claws were sharp but not huge. It was beautiful and graceful as it moved, and when it locked eyes with Lillith, there was a ping, and she became its partner; knowledge filled her mind.

“Oh Gods, what is that?”

“Her name is Prasia, and she’s a dragon.”

“A what? Aren’t those supposed to be mythical creatures made up by man?” Andrew pointed out. “How did you know her name, and that it’s a she and that she’s a dragon?”

“Well, obviously they are real, tiny, but real. I think we locked our minds, and she told me,” Lillith pet Prasia, she loved her partner, partner in what? She was confused, as if a big chunk was missing from her head. Lillith pet Prasia and listened to her purr and make happy noises.

****

Lillith lay in bed, wondering what to do. She let her partner sleep under her chin. You’d think a dragon with scales would be harsh on the skin, but Praisia was so soft; she worried she’d crush her. So, she moved her to her shoulder. Praisia was fine with that.

****

The next day, Lillith had a good night's sleep, thinking it was Praisia's doing. She took a shower with Praisia. Praisia didn't mind the cold water. She brushed her hair, teeth, then got dressed. She then heard a knock at the door. She looked for her mother and father, but they were both gone. Lillith opened the door, and a child she had never seen stood there.

"Can I help you?" Lillith asked the child.

"The elder witch wants you to visit her." Then the child ran away.

"Wait, what?" See the Elder witch? She knew where she lived, so she was definitely going to get Andrew to come with her. She stepped outside and stretched. Lillith walked to Andrew's house down the street, then knocked on the door. Andrew answered half asleep.

"Wake up, get dressed, and look decent, we are going to see the Elder witch. She wishes to see me for some reason. Andrew said hello to Praisia and petted the tiny little dragon, then ran inside and got all fixed up. He went back to the door. "Let's go!"

****

Lillith and Andrew came to the outer edge of town, near the forest, then knocked on the door.

"Come in, my sweets, you must come in!"

They entered.

"Who's this boy? Oh, your love, then he is welcomed. So, I assume you've heard about the evil things taking over the towns? Do you want to know what the evil is?"

Lillith's heart pounded, "Yes, please."

"The evil is a high priest skinwalker, he is the leader, he has the Wendigo that are destroying the world, so it can be dead like them. The Wendigo devour humans alive, and kill and destroy, and then the lower skinwalkers burn everything to the ground and kill the ground and everything around, turning the places into a dead zone. They are planning on killing all of Ethridge, they already killed all of Chella, they want all of Eterial dead, then people will die without the world. You must defeat them."

"Defeat them, all alone?" Lillith's heart pounded stronger, and her forehead almost hummed.

"Oh no no no, you have to find your warriors, the ones that can defeat evil."

"Warriors already tried and failed badly!" Lillith was getting dizzy.

"Oh, come, please sit, my poor dear." The Elder witch cleared her throat. My dear, we need to use magic, a special, ancient magic that is over two hundred years old. You need to find your four warriors; you must look far and wide."

"How the heck am I supposed to find them?" Asked Lillith.

"You will use magic and your dragon; they will fly you everywhere you need to go."

"For one, she is tiny, and two, she is a baby!"

"You shared your love with her, she is an adult now already, right away from hatching. And she can change her size to give you a ride wherever you need to go."

"No way, and how am I to find these warriors?"

"With this crystal, my dear." She handed her the chain, which had a beautiful rose-colored crystal in it.

Lillith looked at Andrew, and he was staring at the Elder witch.

"How do I use it?" Lillith was so lost, but she paid attention.

"You put the crystal to your eyes, then look through it, and then you will see a town glowing. You will tell your dragon to go down, and then you will see this person glowing with a blue aura just like the town, as you look through the crystal."

"How am I to convince these people into fighting?" Lillith was getting frustrated.

"Put this crystal on them, and they will be filled with all that you know." She handed Lillith a clear crystal to use.

Lillith sighed a big sigh. “Okay, I will trust you, but only because this little dragon showed up. She made me a believer.”

“Oh, you must also find a partner for each member; you have to do the same thing you did to get your partner, for them to get their partners.”

“Okay, this is getting exciting. Then we go fight the evil when I collect them all?”

“NO,” The elder witch almost yelled. You must come back to me, your journey has many things you must get to be ready, it’s not going to be easy, but it is worth it to get the world the way it used to be.”

“Okay, I will pack some food......”

“I’m coming with you, no ifs ands or buts.” Andrew announced.

“Yes, your love must go with you for support, he is your one true love, don’t let it go, hold onto it tight.”

“Okay......I was going to say yes, but what you said didn’t make things too awkward.” Lillith giggled and left with the crystals. She wasn’t going to tell her parents, but they would find out when they saved the world.

****

“Okay, we should have everything we need. Are you ready, Andrew?” Lillith smiled at him.

"As ready as I'll ever be." They walked out of town so people wouldn't see them.

Lillith placed Prasia on the ground. "Okay, what now?"

Nothing happened.

"Hmmmm...." She thought, "Please grow."

Just then, Praisia grew to an extremely large dragon. She looked beautiful, with a golden complexion and blue eyes.

"Okay, we are going to climb you now."

Prasia snorted and bent down.

She climbed all the way up, and Andrew followed and sat behind her.

Lillith took out the crystal, one hand holding the crystal, the other hand gripping the hair, then said to Prasia. "Fly." And whoosh, they were off.

Andrew screamed at first, and Lillith laughed at the rush.

About five minutes after rising into the sky, she raised the crystal to her eyes.

Praisia made a calm, small roar. This made Lillith laugh. She really enjoyed being up in the air with Prasia and Andrew.

Andrew held on tight and pressed against Lillith; this felt very nice to her. She was not alone with these two.

****

After riding for what felt like forever, Lillith saw a glowing city and got excited.

“We found the first Town! Isn’t this exciting, Andrew?”

Andrew squeezed Lillith’s waist and laughed. He was just getting used to flying. He found it terrifying to go from ground to sky; he could only imagine how scary it would be to land.

Down they went; Andrew couldn’t help but scream. It was worse than falling! Andrew thought.

They landed outside of town, where they wouldn’t be noticed. Then, climbing off of Praisia, Lillith felt so alive! Andrew almost kissed the ground when he got off, then Praisia shrank and flew onto Lillith’s shoulder. Lillith petted her and thanked her for the ride.

“People are going to notice Prasia,” Andrew said.

“Keep your wings tucked in, okay, Praisia?” Lillith said while petting her, and Prasia chirped and tucked them in. “Good girl.” Lillith praised.

“Oh crap, did we bring money?” Asked Andrew.

“We have so much money, we would be considered royal. I have been saving money since I was five, and I earned a significant amount from various sources, including bets, side jobs, and my allowance. My parents were well off too, you know that. My dad trained men, and my mom sewed dresses; we barely spent anything. I mean they bought the house, so no rent, we just didn’t let people know it, kinda our secret.”

“I’m starving,” Andrew whined.

“Okay, it’s night time, let’s go to the pub, there should be one? They looked down the street, lined with houses made with wood, some falling apart, some fancy, there has got to be a pub! I want steak, RARE!” They walked until they came to the main street with shops, and saw a pub tucked in the middle, between two shops: one for flowers and the other a butcher’s.

“Least you don’t eat like you used to,” he laughed.

“I dunno, I’m pretty hungry, might attack my steak!” She laughed.

“Please no,” Andrew begged.

“Okay, okay, I will eat normal.” She sighed. “Let’s get to that pub, see if we can find the warrior!”

****

"Okay, we are about to go in, I might get stared at because of my long straight red hair, and you with your long blond hair in a braid, not to mention we are strangers, and I have a golden lizard on my shoulder, so be prepared," Lillith warned.

"Don't worry, we will get our food and look around. If they are not in here, we will have to look for a while, but let's get food first."

"Sounds good to me!" Lillith laughed.

The two walked into the pub, went to the bar, and ordered a rare steak and chicken wings. Then, they went to sit in the corner of the bar.

Everyone was staring at them; they were new, so they were probably seen as dangerous.

The waitress brought them the steak and chicken wings; it took less than half an hour to arrive.

"Looks good, thank you." Lillith smiled at the waitress.

"Can I get you two something to drink?"

"What do you have?" Lillith asked.

"We have beer, wine, hard liquor, and a drink called cream soda."

"That sounds great. Both of us would like two large cream sodas?"

"Yes, of course!" She turned and went to the bar.

"Have you ever heard of cream soda?" Lillith asked Andrew.

"Nope, let's dig in!" Andrew said happily.

They started to eat, and oh my goodness, the steak was so juicy. Lillith cut up some for Praisia, and the chicken wings were crispy. They were very pleased.

The waitress brought them two large glasses of pink bubbly stuff.

"Thank you!" They both said.

"No problem." She giggled, then left.

Andrew looked at the drink and wondered.

Lillith picked the cup up and took a drink. "It's good!"

Andrew took a drink. "Oh damn, this is really good!"

They continued to eat and drink.

****

Lillith ate everything, including the fat; her father said, "Never waste a good meal." Giving Praisia little chunks every time she took a bite."So good!"

“Very tasty,” Andrew agreed. The waitress came and took the dishes away.

“Okay, I’m gonna see if they are in the pub,” Lilith said, taking the crystal and looking through it. She scanned the room, and it glowed around a young man with blue hair.

“Oh my God, he’s over there!” Lillith took a deep breath. “Okay, I’m going to go talk to him, you stay here, no matter what,” she told him.

“Okay, I promise,” Andrew replied.

Lillith stood up and walked over to the blue-haired man, then sat next to him.

“Can we talk?” She asked him.

The man looked up, no older than Andrew’s age.

“You know redheads are bad luck, should I listen?” He joked with her.

“Yes, I would hope you would listen, it’s very important.”

“Okay, spill it.” He laughed.

“Okay, we are in search of warriors, four to be exact, you are number one, we need to save the world,” Lillith said, maybe a little too blunt.

“Ha, what?” The man looked at her. “You know, I’m going to listen to you, you have the most beautiful purple eyes, they match that mark on your forehead.”

“I’m not here to flirt, I’m here to find my warriors.”

“Okay, so how am I supposed to believe this? You could be insane.”

Lillith took out the clear crystal. “Touch this and you will learn like I did.”

“You probably want to steal my soul. How can I trust you? You could be a witch.”

“Trust me, I know you don’t know me, but I’m being honest, I wouldn’t hurt you.” Lillith pleaded.

“Okay, I will, but only because you are beautiful.” He laughed, then placed his hand on the clear crystal. As soon as he touched the crystal, his eyes widened, then rolled back into his head, and he fell off his stool, remaining in shock for a minute before closing his eyes and stopping all movement.

“What the hell did you do to Dom!’ An old man with no hair yelled at Lillith, then attacked her.

She blocked and dodged all his punches. “I don’t want to hurt you! Please stop!”

Dom sat up and rubbed his eyes. "Hey, stop it! Who goes to hit a woman anyway? What kind of coward are you, Mike!"

Lillith looked down at Dom. "Are you okay?"

"Yeah, just a ton of info shocked my system. I need to go. Meet me here tomorrow, we will talk. I live across the street. If you need anything, don't be afraid to ask." Dom got up and left the bar, everyone in the bar just stared at Dom as he was leaving.

Lillith walked over to Andrew. "Thanks for not interfering."

"I knew you could handle it." Andrew smiled.

"Okay, we have to find a partner for Dom, so he can believe. Do you want to come?" Lillith asked Andrew.

"Yes, I do." He smiled. "But it's dark out." He was worried.

"Don't worry, the stars will make it not so dark.

"Okay, hun."

****

They walked around in the forest, but nothing had been found yet. Lillith started to get worried, but she knew she would find it, no matter how long it took. They arrived at a small, desert-like part of the forest. Dunes. She started to climb up one, and her birthmark started to warm up. "We are on the right path!" Lillith exclaimed.

"I knew you could do it, Lillith!" Andrew exclaimed.

She followed the warmth until she got to a small waterfall in a pond. Her forehead began to burn.

"It's gotta be here!" She looked in the water and saw an egg at the bottom; it looked just like hers, with markings, but it was a beautiful blue. She carefully picked it up. The same size as her egg was.

"It's beautiful," said Andrew.

"Yes, and now we have to deliver it right away, so it can hatch!" Lillith was relieved.

****

The two knocked on Dom's door.

"Just a second!" Dom shouted.

Lillith was happy; this was really working out for the better.

Dom opened the door and sighed. "Yes?"

"Oh, I found your partner!" She held out the beautiful blue egg.

Dom looked at it and smiled, Oh yes, my partner!" I was afraid I had to find it on my own." He confessed.

"No, it's part of my job as the leader."

"Thank you." He carefully took the egg.

"Now send all your love into the egg and it will hatch when it's ready." Lillith smiled. "This is my partner Praisia." She showed Dom.

"Oh wow, a golden lizard, oh wait, no, that is a dragon." He remembered.

"Yup!" Lillith smiled.

"Okay, I'm going to bed. I need to think about things."

"Okay, good night, Dom," both Lillith and Andrew said in unison.

"Good night." Dom closed the door.

Lillith and Andrew walked towards the inn at the end of Main Street.

"What if he refuses?" Andrew asked.

"He won't, I just know." She giggled.

****

Dom took the egg and laid it on a warm pillow next to him.

"I'm not sure how, but I will send all my love to you; you are very important to me." He then closed his eyes and fell asleep. He woke up out of nowhere, in the middle of the night, and noticed his egg was hatching. It exploded out of the shell. A beautiful Blue dragon with golden eyes popped out; it had no horns. The dragon hopped on Dom's chest and stared into his eyes as Dom stared into the dragon's eyes.

"Oh wow, you say you are a dragon? And your name is Umi. Beautiful just like you. My name is Dominic, but I go by Dom," Umi chirped happily and nuzzled Dom's face. He was very happy; this just proved he was needed. However, the problem was that he had never fought with a sword in his entire life. How can he be a warrior and even fight the wendigo? He frowned and asked Umi if he could go back to sleep. She nodded and perched on his shoulder. He passed out.

****

Lillith woke up, cuddling with Andrew. It felt so warm and safe in his arms. She had been in love with him since ten; he was her best friend and fiancé. She stayed still and relaxed while Andrew slept, and he looked so innocent sleeping. He was her perfect match. She wondered how Dom was; she wondered if his egg had hatched yet. And what would it be? Could it be a dragon? Who knows, whatever it is, it will be special.

Andrew stirred, but didn't wake. Lillith thought that was adorable. He was so handsome, with the most beautiful blue eyes and long, golden hair that he normally put in a braid. She liked that. She had long straight red hair, with lilac eyes. She felt they made a good match.

She couldn't see herself with anyone else, not even the most handsome boy to all the other girls. His name was Argon, and he was supposedly a dream guy. He asked Lillith out, and she told him she was already dating Andrew. And he had the balls to say, "Dump that loser for me, I will treat you like a queen." Lillith told him not to talk about her fiancé like that ever again, or she would skin him alive. No one was as good-looking as her man. She hugged Andrew tighter.

Praisia stretched and walked onto Lillith's chest. Lillith looked down at her.

"Good morning, Praisia," she whispered.

The little dragon cooed at her and gave her kisses.

"I love you, too, Praisia."

Praisia got annoyed because Andrew was taking his time to wake up. She flew up to his face and made a tiny roar.

"Wah! Prasia?" Andrew woke up shocked.

"Praisia, that was not nice to poor Andrew," Lillith whispered, trying to hide her giggles.

Andrew sat up while Prasia climbed onto his shoulder.

"Okay, I'm up now!"

"We should get dressed and go back to the pub and have breakfast!" Lillith advised.

"Sounds good!" Andrew agreed.

Praisia squeaked and flew to Lillith's shoulder.

****

Lillith and Andrew entered the pub.

"Do you serve breakfast?" Andrew asked.

"Yes, we do, what would you like?"

"Two eggs, four sausages, and a cream soda, oh, and bread, both will have the same."

"Coming right up!"

They sat back in the corner.

"I wonder when he will come, if he does," Andrew pondered.

"Who knows, but we will wait here all day and night if we have to," Lillith stated.

"Okay." Andrew agreed.

****

Dom and Umi entered the pub at noon. He looked around and saw the two strangers he had met and brought Umi from. He walked over to them; they were eating a salad. Lillith looked over and smiled at him. "You came, come, sit down."

Dom sat down with them, not looking too happy, but Umi chirped in his ear; she was so little. "What a beautiful dragon!'" Lillith looked at her. "Oh, tell your dragon to hide their wings."

"She is, isn't she? Her name is Umi. Oh ok, Umi, could you hide your wings?" Dom asked Umi.

"Oh, she's so pretty! Lillith went to pet her, and Prasia jumped down and smelled the new dragon. They squeaked and they rubbed on each other.

"Can I pet her now, Prasia?" Lillith asked, Prasia nodded, so Lillith started to pet the beautiful blue dragon.

"Anyways, my name is Andrew, and this is Lillith," Andrew introduced them.

"I'm Dom, nice to meet you."

"What is the name of this town, by the way?"

"Oh, it's Liel."

“Are you going to join us?” Asked Lillith, who looked excited.

“You see, I want to, but I am not a warrior....I don’t know how to fight with swords.” Dom admitted.

“Ohhhh....Well, you’re in luck. Andrew and I are the best swordsmen in the land! We can train you! We will help you out.” Lillith smiled.

“I hope I’m worth the trouble.” Dom said, then Umi climbed up to Dom’s shoulder and cuddled his face.”

“Oh, you bet you are worth it, you are a part of our team, we need you,” Lillith admitted.

Ok, we will eat and then we will practise. They ordered food and three cream sodas.

****

They rented two dull swords from the weapons shop and practised. Dom was actually a very fast learner and used the skills they taught him against them. He wasn’t even close to being a professional, but Dom knew now how to attack and protect himself. That was good so far; he would need to continue practicing, but he was getting better. He swung his sword like he had done it all his life, and his form was very good.

“I think we should go to bed, it’s getting late, and I want to leave semi-early, maybe lunch time,” said Lillith.

"Okay!" Dom said excitedly.

Just as they were leaving, both Praisia and Umi sneezed and coughed.

"Oh wow!" Lillith cradled Praisia in her arms, and Umi lay down in Dom's arms.

They all looked at each other. "What's going on?"

CHAPTER 3 –
THE DRAGON'S BOND

CHAPTER 3

THE DARGON'S BOND

Lillith looked at Dom. "Do you have a witch in town?"

"No, but one in the woods, I used to bring her firewood, you think she will know what to do?"

"Well, it was an Elder witch that set me out on this journey, so why not?"

"Good point, let's go, it's dark though, so we have to be careful, not to mention if she's up."

They went to the forest to get to the witch's house.

****

They finally arrived at the witch's house/hut.

Dom knocked on the door and waited as Umi cooed sadly.

"I hope she can help." Lillith looked down at her poor Prasia. "She is getting weaker....." Tears began to well up in Lillith's eyes.

The old witch opened the door. "Welcome!" and let them in.

“My name is Lillith, this is Andrew, and that’s Dom.”

“Nice to finally meet you, Lillith. So your partners are getting sick? I’m so glad you came to me, the thing that would cure them, I have been collecting just for the occasion.”

“You knew we were coming?” Asked Andrew.

“Well, I had a gut feeling, always follow your gut, let me get you the big bag that I collected.” She smiled and went into the back. “You see, they are getting sick because the world itself is dying, from the evil creatures that are destroying the land,” explained the old witch.

“How fast are they going?” Asked Lillith.

“I’d say they got, from what I heard, four towns,” Dom replied.

“Oh no..” Lillith almost cried. “Was she failing?” she thought.

“It’s okay, dear, you are not failing,” the witch calmed.

“How did you know what I was thinking?” asked Lillith.

“Lucky guess.” The witch giggled.

She brought the bag to Lillith, and Lillith put the bag into her satchel.

Then, they watched the dragons eat the nuts, and each looked like they were normal again.

“Thank you so much,” Lillith said loudly. She went to pick up Prasia, but Prasia shuffled and looked like she was breathing hard.

“What’s wrong with her?” Lillith demanded.

“Well, I’m guessing your friend over there,” She pointed to Andrew.”Has been giving this little dragon lots of love. She’s giving birth, now watch the rarest thing you can see these days.”

Praisia pushed and popped out a silver egg. Praisia breathed calmer, then stared at Andrew.

“Praisia, are you giving me your baby?” Asked Andrew.

Praisia nodded and pushed it towards him.

Andrew picked up the little silver egg and held it close to his heart.

“Thank you, Praisia, I will love it forever and treat it as good as I treat you.”

Prasia jumped up onto Andrew and licked his cheek, then flew to Lillith’s shoulder and kissed her.

The old witch laughed. “She has got quite the personality, she must love you both, but I need to warn you, Andrew you are not a special warrior, so please stay out of the fights, keep yourself safe, we don’t want you to die, it is very noble of you to come with your love, but you could get hurt, so please, don't fight or your future will be empty."

"I will leave the fighting to these guys, I will just support my Lillith."

Lillith smiled and kissed Andrew.

"How did she get pregnant?" Lillith wondered.

"It's rare, but when a dragon gets a lot of positive energy, it is possible to make a baby out of that love, but hundreds of years ago, it would happen once in a millennium. Wow, I knew you guys were coming, but never would I have guessed I'd be lucky to see a dragon give birth!!" The old witch laughed and clapped.

"Wait, how did you know?" Asked Dom.

"Oh, I've known since I was a younger witch, the elder witch has told many of the witches that you may need our help, so if you need help, look for us, we are here to help you."

All three were shocked that all this had been predicted. "Don't be so shocked, all the witches in every town have been told to look out for you guys in 18 years, it's a prophecy passed down from the first witches."

All three looked at each other, dumbfounded.

They eventually left. Dom went to his house with Umi chirping happily, and Lillith and Andrew went to their room. They got ready for bed and then lay down. Andrew put the egg on a pillow in between the two, then Praisia curled up around her baby's egg. They kissed and then went to sleep.

****

Lillith woke up because Praisia started to chirp. The silver egg was moving. Lillith woke up Andrew.

"Yeah, hun?" Andrew answered Lillith.

"Your egg is hatching!" Lillith said loudly.

Andrew sat up and watched the egg; it rocked a few times, then, poof, it burst out like the shell was nothing. It was a silver dragon with beautiful purple eyes and four horns on its head. It then cooed and climbed up Andrew's shirt, staring into his eyes, and he stared back at it.

"Oh, it's a boy dragon, and... you have to be kidding??" Laughed Andrew.

The dragon insisted that it was his name.

"Okay, his name is Bob."

"Nice name," Lillith said with a huge smile on her face.

Bob then crawled to Andrew's shoulder and squeaked.

“Ok, let’s get back to bed. “

Bob purred and lay down, and they all went back to bed.

****

They woke up and got ready.

“You know, we should buy swords and dull swords to practise with.” Lillith decided.

“Actually, that sounds like a good idea, let’s go to the shop.” Andrew agreed.

So they packed their stuff and went to the sword and shield shop.

“Um, I’d like to buy 3 of your best swords and two dull swords,” Lillith said.

The man let them in. “My best swords are right here, and the dull swords are over there.” He pointed to the back of the small shop.

Lillith looked at the swords; they were beautiful. There were four to choose from, and Lillith picked out the three she felt were the best, then took two dull swords.

“We will take these five swords.”

"Beautiful choices, they all come with scabbards, so it's actually a good deal."

They paid a pretty high price; each sword, excluding the blunt swords, cost fifty silver. However, they were a good size, had good weight, and were balanced just right.

"Thank you," they both said and left. They then went to Dom's house and knocked on the door.

Dom answered, still sleepy. It was noon, but it looked like he had had a hard time sleeping last night.

"Looks like you didn't sleep well?" Lillith said.

"Nope, I slept well, just too much, are we leaving?" Dom asked.

"We are going to eat first, then go. We got you a sword, just in case, and two blunt swords. So we can practise." Lillith said happily.

"Oh, my own sword, that's amazing!"

Lillith handed it to Dom, he looked at it, and smiled." I feel so much stronger." He laughed.

Umi chirped excitedly.

They all headed to the pub across the street, and they all had a big breakfast/lunch. But once they were done. Dom and Andrew practised fighting for an hour. Dom wasn't half bad, but he needed to practise. He was good at blocking and attacking, but there was much more he needed to learn. But pretty good for a beginner.

"Okay, let's look for number two!!" Lillith exclaimed.

They all left the town and put their partners down; all three grew huge. They all climbed on the backs of their partners, and they all flew up. Both Andrew and Dom screamed. Lillith laughed.

****

Lillith looked for hours, and their partners began to tire out, so Lillith found a small town and had the dragons fly down to it. They set down softly in a field next to the town.

"Ok, guys, our partners need a break, so we might as well find a pub in town and let them relax, then fit in some sparring a little, then leave again to search."

"Sounds good to me!" laughed Dom.

"No drinking beer, okay, or alcohol! We need to stay alert."

"Understood." Both men agreed.

So they walked into town, in two lines on the street. They wandered around, most of the houses were attached to each other, but some were separate. They found the main street, and there were plenty of stores, but all they were interested in was the pub that was on the corner. It was dark now, and all six of them were starving. The pub looked kinda run down, but that was probably because of the weather. They all entered, and everyone looked at the three strangers, who had lizards on their shoulders. They had to look out of place to the people in the pub. But oh well, they all sat down and looked at the menu.

"Wow, this place has a menu!" Lillith squealed.

The waitress came to the table and ordered three cream sodas to go with their meals.

All of their partners were sleeping on their shoulders.

"Aww, they are so cute sleeping." Lillith praised.

"I wonder why we haven't found the second yet, but this is a huge continent, bigger than the other continent, nearly half the world, the continent across the sea isn't even half as big, so it may take some time. We just don't have time to waste, but we can't search forever, our partners need breaks," says Andrew.

"Yes, they do, and we will stop all over if we have to for them." Lillith sighed. "I'm just worried about the people being killed and the world dying before we get everyone."

"I know, hun, but we can only do so much till we get all of us together." Andrew comforted Lillith.

"Yeah, you are right, sorry, just scared we are taking too long."

“We aren’t, we are in the lead, I feel it,” Dom said happily.

“I hope so....” Lillith was still worried.

The food came and they ate it all, sharing it with their partners.

“Well, it’s late, we should get rooms at the inn next door,” Dom suggested.

“I guess we don’t have a choice, can’t see in the dark. Lillith admitted.

****

They woke up early. Lillith sparred with Dom while it was still a little dark outside. As soon as the sun was shining, they were off again. A couple of hours later, Lillith saw a light coming from a city.

“We found the city! Let’s go down!” Lillith screamed happily and told the group to land next to the town.

“Okay, let’s go to the pub, because it’s where we have had the most luck finding our fighters. Plus, I’m hungry.” Andrew pointed out.

The group entered the city, which seemed like an old town; every house was stone, and some even had gardens. They wandered until they reached the centre, where there were so many shops! Like anything you could want, there was a store for it. They almost passed the pub; they noted the Inn was on the other block.

They walked in looking like weirdos with their partners on their shoulders.

As Lillith entered last, a rude man with short silver hair nudged Lillith out of the way as he entered the pub, almost knocking her over.

“Oh my Gods! How rude!”

“Hey, that was very rude!” Andrew growled, “Are you okay, Lillith?”

“Well, she shouldn’t have gotten in my damn way!” Says the man, who goes right to the bar.

“Yeah, I’m okay,” Lillith said, glaring at the rude man.

Let’s get something to eat, forget about rude people.” Dom suggests.

“Okay, I’m starving!” Smiles Lillith.

They eat their food and drink their cream soda.

Okay, let’s see if the person is in here,” Lillith takes out the crystal and looks around, then stops, mouth hanging open.

“What?” Asks Andrew.

“It’s the rude man....”

"Oh no...." Says Dom. "He is not going to want to join for sure."

"Let me handle this; no one interferes." Lillith walked over to the silver-haired rude man. "Can I buy you a drink?" Then sat next to him.

"Damn straight!" Then smiles.

"Sorry, I got in your way earlier." Lillith apologised. She then noticed that he had a wooden staff next to him.

"You should be, but it's ok, you bought me a drink." He took a sip. "So what do you want? I know you wouldn't do this if you didn't want something."

"Well, it would be a lot easier if you just touch this crystal...." Lillith took it out and placed it between the two.

"Is this a joke? I am a travelling witch. What if this is a trap? I don't need a curse or lose my powers because I bumped into another travelling witch herself."

"It's not a trick, it's an opportunity to help save the world. And I'm not a travelling witch."

"How so?" He asked.

"Well, haven't you heard about the towns being destroyed by unbelievable evil creatures? The whole other continent is destroyed?"

“Yes, but how can I stop them? Many travelling witches have gone up against it and got eaten alive or just killed, how could I help?” exclaimed the witch.

“Touch this crystal, and you will be filled in,” explained Lillith.

“Fine, but if something bad happens, you are going to suffer like no one has ever.” The man slowly reached over and touched the clear crystal. His whole body shuddered, his eyes went back into his head, and then he flopped onto the bar.

Lillith put her hand on his back.

“Don’t… touch me.” He mumbled.

“Do you understand now? Will you join us?” Lillith asked the man.

“Looks like destiny has called me, I don’t have a choice.” He sat up and sighed. He stood up.

“Come sit with us and get to know us, and we can get to know you,” said Lillith

“I don’t do touchy-feel good kind of things,” answered Jack.

“Well, you will be saving the world, so you better get used to being a good person.”

He glared at Lillith, then rolled his eyes. “I guess.....”

They both walked to the table and sat with the group.

“So, dragons exist, and I get a partner?”

“Yes! I have to find your partner right after we are done chatting.” They walked to the table where Andrew and Dom were sitting.

“Can I come with you?” Asked the man who could only be 25 at the most. “I mean to get my partner?”

“Yes, you can, what’s your name? My name is Lillith. This is my fiancé, Andrew, and this is Dom. You will be our second member! Two more and we can hopefully start the fight? But we dunno, The Elder witch said there’s lots to do.”

“My name is Jack, and I’m a travelling witch.” He knocked his staff on the ground.

“That is so cool!” Dom praised.

“How long have you been practising?” Lillith asked.

“I was born into it, I’m not the best, but I’m not the worst either.” Jack laughed.

“Wow, your eyes are a beautiful green, forest green, so pretty.” Lillith pointed out.

“They are not pretty, they are sexy.” Jack spat.

“Oh yea, so sexy.” Lillith laughed.

“Are you making fun of me?” Jack looked pissed.

“A little.” She laughed harder.

“Well, you noticed, so they must be.” He glared at Lillith.

“Okay, let’s go find your partner!” Lillith announced.

“Okay, I guess,” Jack said, hiding his excitement.

“Wait, what is the name of this town?” Asked Andrew

“It’s called DeVille.”

“Okay.” Andrew nodded.

They all left and headed for the forest.

“It might take a bit of wandering to find your partner.” Admitted Lillith. This man made her doubt herself.

“That’s fine,” Jack said calmly.

They wandered for half an hour until they hit the creek.

“I can feel it!” Lillith laughed. The mark on her forehead got warmer until it burned. "It's got to be around here somewhere, my forehead is burning!" She saw a hole in the ground, so Lillith bravely put her hand into it, and she found something. It was deep in the hole. It was a perfect circle and sage green.

Jack's eyes widened, and he actually smiled.

"Here is your partner, you have to fill it with love, and it will hatch."

"Thank you!" Jack cradled it in his hand. "Let's go back into town, it's late."

"Yes, we should get rooms at the inn." Dom thought.

"I will meet you guys in the pub in the morning," Jack told them.

"Sounds like a plan." Lillith laughed.

~~~~~~~~~~

Jack entered his home, looking at the round lil egg, he loved it instantly. He lay down and placed the egg next to him. He fell asleep watching it. He woke up out of nowhere and noticed the egg shaking. Jack held his breath. It burst open, and out popped a dragon. The sage green dragon was a unique kind of dragon, characterized by its fur and scales, as well as a dog-like face. The face had a darker green mouth and ears, and its eyes were silver. It also had two horns on its head. He purred and locked eyes with Jack and told Jack his name was Merlo. He cuddled Merlo. They both fell asleep.
~~~~~~~~~~

~~~~~~~~~~~~~

They met at the pub early in the morning. They waited while eating breakfast.

Jack walked in with the interesting dragon on his shoulder. He sat next to Lillith.

"What on earth is that?" Asked Andrew.

"It's a luck dragon."

"I think it's beautiful! Lillith said, reaching over to pet the dog's head of the dragon. All the dragons went up to the new dragon and greeted him.

"His name is Merlo."

"What a nice name. I love his sage color," said Dom.

"I like it as well," Jack said calmly.

"Do you know how to fight? And are you really into that staff, cuz I'd like to buy you a better, stronger one." Lillith asked.

Actually, I could use a stronger one.....But I don't fight much...." He looked at Lillith. "Is that a problem?"

"We are going to have to train you then," Lillith announced. "First,
~~~~~~~~~~~~~

let's get you a new staff, next it's training. After breakfast." Lillith nibbled on her toast.

"Sounds like a plan." Jack smiled at Lillith.

****

They walked down the street until they found a weapons shop.

"Choose the strongest, ok?" Lillith told Jack.

"Okay, if you wanna pay that much." Jack walked into the shop. "Shopkeeper, what's your strongest staff?"

"Well, let me see, this one is made very strong, and will get you through any fight!" He handed it to Jack, and Jack rolled it around in his hand. On both ends of the staph were sharp points, perfect!

"I really like this one." Stated Jack.

"Oh, I dunno if you could afford it, it's mighty expensive," said the shopkeeper.

"I will pay for it," Lillith said, and she handed him the money; then they left.

"Wow, I feel strong." He started to whip it around and lunge at nothing.

"We need to train now, you will train with me, and Dom will train with Andrew," said Lillith.

"Woah, who made you boss?" Jack asked defiantly.

"The Head Witch made me leader," Lillith smirked.

"I don't want to fight a girl......" Jack said angrily.

Andrew laughed.

"You will fight and lose, and then lose again, until you can protect yourself!" Ordered Lillith.

"Oh, ok, I guess, but if I hurt you, no whining."

Lillith handed Andrew and Dom the dull swords, and Lillith used the sharp one, only because they had only two dull swords.

****

After a few hours of training, they were all exhausted, but at least now Lillith knew Jack could handle a fight and defend himself, same with Dom, he was improving so much."

"Holy crow, can she fight!" Jack announced, almost out of energy.

"That's why I laughed when you said you don't want to fight a girl," laughed Andrew.

“Let’s go to the pub, I’m starving and need to rest.” Said Jack.

“No drinking! Try cream soda, it’s yummy.”

“Fine, no drinking....” Jack pouted.

****

They ate, laughed, and shared with their partners. It was great times, like they were old friends having a great meal together. When they were done, they chatted over cream soda until they were revitalised and ready to be on their way. They left the pub, and as soon as they put their partners on the ground to fly, Lillith paused.

“Stop, everyone! I sense something or things coming, and they feel evil!”

All of them turned around, and Lillith turned and told the partners to hide! They did as they were told, but they were worried about their partners.

“Get in stance, they are coming, get in a circle, I don’t know which side they are coming from!”

Then they could hear the growling.

“Everyone in the town stays inside or faces death!” Lillith screamed, and all the people on the streets went indoors.

Then out of the forest showed the twisted evil faces, coming from the east, they looked starved and skin and bones, not to mention rotting, and smelled of death. It moaned. "hungry.....HUNGRY!"

Lillith yelled, and they all faced the direction. The bastard High Priest skinwalker must have sent some all over the continent, just to stop them. He knew they were coming and wanted them killed before they even had a chance. She trusted her team; they could fight a good fight and win. There were only three.

The first one lunged at Lillith, but Lillith was ready and kicked it away, then sliced hard on its shoulder. These things didn't give up till they were dead. The other two lunged, attacking Dom and Jack. Jack held his own, hitting and stabbing the one while Dom was slicing and striking the other. Lillith took a step back, while backing up, the one swiped at her with its razor claws and sliced her arm up good.

"Ahhhh DAMMIT!" She swung so hard she took the head off. They were all getting hurt, but they were holding their own. The partners saw their humans getting hurt. Praisia jumped in, turning huge. Praisia bit one in half; the other wendigo saw this and jumped on Praisia, scratching her up. She spat out the disgusting, tasting creature that was now crawling with only its arms. "No Prasia!" She jumped on the wendigo and pulled it away from Praisia, then stabbed it in the heart. Andrew grabbed the other and chopped its head clear off. Then Jack plunged his staff into the skull of the one in half pathetically crawling towards Lillith.

They were all covered in wounds, even Praisia was, but hers was more like scratches, some deep, some shallow. They went to the inn and asked the innkeeper where the witch would be. He told them to go to the edge of town on the western side, where her hut would be.

****

They went to the witch's hut right away to help with their wounds. Lillith carried poor Praisia in her uninjured arm. Poor thing was scratched up.

"Witch, please help us....." Lillith said weakly, then blacked out from blood loss on the deep wound on her arm. She fell to her side so she wouldn't hurt Praisia. Andrew ran to her, lifted her up, and carried her to the hut, then knocked on the hut's door.

The witch came to the door and pouted. "I had a feeling you guys would be hurt, felt those evil creatures and knew you were in town, come in, come in, I will fix you all right up."

Andrew laid Lillith on the cot, with Praisia guarding Lillith with her little life.

"Ok, the worst is first, hmm..." She looked at Lillith's arm and knew that was the worst. She tied a rope around the bicep, then grabbed a large jar of healing herbs jelly and rubbed it on. Then, she took out a needle and thread from the counter and sewed every cut up, including the vein in her arm that was leaking. She then sewed the rest of it up, applied more salve, and bandaged it. "She has the marking on her forehead; she will heal in no time, no worries, little one." She patted the little golden dragon on the head.

special ointment on Praisia to help her wounds heal quickly. It only worked on magical beasts, which they didn't have in this world until now, but she had it. Dom had a head injury, needed stitches. Andrew had a cut to the neck, but it wasn't deep, so just some salve. Jack had a deep cut on his thigh, stitches. They were all beaten up pretty badly, but it wasn't a fair fight, only two of them were worthy fighters, the other two were juniors, this was a cruel joke to the Head skinwalker no doubt knew they were getting ready to stop him, but he doesn't know where they were, prob sent many troops, Gods be with the people who run into the others. At least this means he is scared they are getting together. Scared they are getting stronger.

"Here, little one, to get the nasty taste of evil out of your mouth." She handed Praisia some berries and mint leaves, and it really did get the nasty taste out. Praisia purred.

****

Lillith woke up in the Inn room, groggy and sore, then sat bolt up looking for Praisia. Praisia coed at Lillith. She was on her shoulder.

"Praisia, you have to trust me, if I say hide, that means stay there, we would have won, just be more damaged, we can't take the risk of losing you, so promise me you won't interfere unless I call to you, and I will if I need it, I promise."

Praisia pouted as much as a dragon could pout and rubbed up against Lillith's cheek.

"No, I'm not mad at you, you are just very, very important to me, and look at all these scratches on my little one." She snuggled her Praisia. Andrew hugged Lillith; he had tears in his eyes. "I thought you bled out and you were going to die..... The witch says you are to take it easy today, but said you are different from us, that you are stronger and can take a beating and keep on ticking. But seeing you like that made me scared. What would I do if you were gone? You are my whole life, my whole world. I'd be lost without you. We have to be more careful, maybe carry shields, so we don't get hurt like that again?" Andrew asked.

"That's a good idea! We will get some tomorrow, but Jack has to use both hands, so we only need to get three." She wiped away his tears then kissed him softly. "We will make it, we will win."

Bob flew to Lillith's shoulder and rubbed against his mom.

"Aww, these guys are too cute." Lillith giggled.

"What time is it?"

"It's supper time, are you ok to go to the pub? I bet the others are there; they are worried about you, but you didn't hear that from me," laughed Andrew, wiping tears from his eyes.

"I'm so hungry," Lillith confessed. Lillith stood up, but down she fell back to the bed. "Ugh, a bit dizzy, but I will be fine." She smiled at Andrew, who smiled back.

"Just take it easy. " He wrapped his arm around Lillith's waist, and they were off. Bob jumped onto Andrew's shoulder and they left for the pub.

****

They entered the pub, and everyone stared at her, but she ignored them and focused on the two smiling men sitting in the corner.

“We knew Lillith couldn’t miss a meal, so we waited,” Dom said.

“Yea, I’m pretty hungry.” Admitted Lillith.

The waitress came over to the table with a smile on her face and took the order, which included four cream sodas.

“Okay, while we wait, we can’t go off for the next few days, we need to heal, but after 2 days we are off to find our other team mates, deal?” Lillith announced, then looked at them all to see if anyone protested; none did. “Aww come on, someone protest, I don’t want to relax for two days.....” She looked down at her hands.

“You need the rest the most!” Andrew pointed out.

“I know.” Lillith frowned.

“Look, I know we have to hurry, but this attack was shocking, and we have to be in top health. I think two days isn’t long enough, but that’s about all we can afford, so two days is good enough.” The waitress brought them their cream sodas and then informed them that their food should be ready in five to ten minutes. Then left.

“Okay, two days it is.”

CHAPTER 4

THE CALL OF THE ELDER WITCH

"What a boring couple of days," Lillith thought.

No sparring for her, no anything! Just lie back and relax, they say. Ugh, I just want to practise fighting! Getting dirty and sweating my brains out! Though there were some good times, playing with everyone's partners, watching them play, cuddling her sweet, caring Praisia. Cuddling Andrew, the guys are sparring too. It was wonderful, though. She and her fiancé, Andrew, would cuddle in between meals. The guys practised their fighting skills. Then just hold each other, forgetting all the problems in the world, and just loving her. Andrew has deep conversations with her new friends and finds out that Jack is not always a jerk! He is very loving to his partner Merlo. Yes, he can be an ass, but he also can be a great guy, and he hides it well. Finding out Dom's hair is naturally blue! But all Lillith wanted to do was collect the rest of her group so they could get to the next task, and hope that there was only one more task, but she had a feeling there was more. How are we going to beat the army of wendigos when we had trouble with just three? Praisia's already healed up, can't tell she has even been scratched. Her arm no longer hurts to move. They also had some fine dining. The pub was great, and since they left, all the pubs have great food. But today we continue on our journey to find the last two people; this continent is too big! But they searched a quarter of it, so if we are lucky and the Gods are on our side, they will be close by.

"Hun, what are you doing? We are done sparring, we are about to leave, get your butt in gear!" Andrew poked fun at Lillith.

"Yay, are you ready, Praisia?" Lillith asked her partner.

Praisia chirped happily.

They all left the town and entered an open field, where they placed their partners on the ground. They grew. They all climbed on each partner and were off. Lillith took out the pinkish crystal and watched below and in front, then side to side.

****

They flew for at least five hours, and they could all tell their partners were getting tired.

"Okay, guys," Lillith yelled so they all could hear. "Let's go down

to that town down there, so our partners can take a breather."

They all flew down, landed softly. Andrew and Dom were now used to flying. Thank the Gods for that. The partners shrank and flew onto their partners' shoulders, hiding their wings. Most of them fell asleep except for Praisia.

"Praisia, please rest and get some sleep, we need to fly again after dinner," Lillith told her.

Praisia huffed, then closed her eyes, and within a minute, she was out cold. Lillith giggled.

The streets were nice and neat around the first few blocks, though the houses were designed pretty much the same. It took them a little while, but they found the street with the shops.

“Oh!!!” Lillith exclaimed as they walked past a weapon shop. “We need to get three shields. Unfortunately, you can’t get one, Jack, you don’t have a third arm.”

“I use my staph to stop attacks, thank you very much,” Jack said snottily.

“Sheesh, Jack, you can be such a jerk. Anyways, let’s get them now!” Lillith ran to the shop. “Shopkeeper?” Lillith called to.

The shopkeeper rolled his eyes. “Yes, what do you think you need?”

“Hmph!” Lillith thought this guy was a snob! “Your most expensive and best shields, we need three, please and thank you.”

“Let’s see, are you sure you can afford it?” The shopkeeper asked, annoyed with her.

“Yes, please show them?” The shopkeeper dug around, then went to the back. “The best I keep in the back, just in case I get robbed. I was going to sell you crappy ones, but that cute little lizard sleeping on your shoulder made me smile, and I had a change of heart.”

“Oh, thank you.” Lillith smiled.

He brought out three beautiful metal shields. “Now these three shields are special; the metal has been folded and hammered over one hundred times, so they are excellent shields, but very expensive. three hundred for each.” The shopkeeper smiled.

“Okay, let me get my purse out.” She withdrew it from her chest, then took out a wad of hundreds, and there were multiple wads in that purse. The shopkeeper noticed and almost had a heart attack.

“Here you go, nine hundred.” She placed the purse back into her chest plate.

“Well, here you go.” The shopkeeper said, still looking at where she carried her purse.

The two guys grabbed one, and then Lillith grabbed hers. “Thank you, kind sir!”

They walked away, but the shopkeeper thought about how much money this short girl had. She didn’t deserve that money. He did, and he was going to get it, then live happily ever after. He just had to get the right men to help.

“Wow, these are so light and pretty.” Lillith laughed as they walked towards the pub for lunch.

****

After lunch, they talked about nothing, just pretty much waiting for their partners to rest and be rejuvenated.

“Okay, one more cream soda, then bathroom break, then we go relax by the big tree beside the pub, sound good?” Asked Lillith.

“Sounds good.” Dom cheered. They chugged down their drinks and all burped in unison, then laughed at each other as pub goers stared at them all. They all went to the bathroom, where they all had to go pretty badly.

****

“Ahhhhhhh, I feel much better now.” Lillith laughed.

“Yes, I feel much better as well.” Admitted Andrew.

“I FEEL LIKE I LOST TEN POUNDS!!!” Dom bellowed for the world to know.

“I feel lighter, too, but I’m not that stupid to yell it for the world to hear,” Jack said snidely, thinking Dom was disgusting.

“Okay, let’s relax..... I will give them one more hour to sleep, then we have to get on our way.” Lillith told her friends.

They all lay back on the tree. They circled the whole tree trunk, with enough room for each of their backs to lean on.

****

The shop owner had two very large men with him. "Look, over there! There is the redhead who has all the money, let's get her!" They ran yelling at her with weapons drawn.

Lillith jumped with shock at an oncoming attack. She looked at the men and laughed.

"You shouldn't have yelled and let me know you were coming!" Lillith jumped up, drew her sword, and assumed her father's stance.

Dom yelped and was ready to attack, but Andrew told him to hold back; she could handle these three grunts.

She took off her shield, and she was ready.

They all attacked at once. Lillith blocked the left two with the shield, then hit the sword on the right with hers.

"Are you sure you want to continue?" Lillith taunted.

"Why wouldn't we, your friends won't even help you." The shopkeeper laughed.

"I don't want to hurt you bastards!" Lillith pushed them away with her shield, then kicked the obviously untrained swordsman in the nuts so hard she knew he would be out.

The shopkeeper fell to his knees and fell flat on his face, making painful moaning noises.

“Would you like to continue?” Lillith asked the last two men holding swords, because they weren't even holding the swords properly.

“We want your money!” The tall man yelled, then swung the sword down, trying to cut her head in half. She easily blocked it, then twisted her sword around his sword and flung it far away, then smashed him on the head hard with the side of her sword. The man fell over unconscious.

“You give up yet?” Lillith asked, annoyed.

The man dropped his sword and ran into the woods until she couldn't see him anymore.

Praisia purred, now awake and proud of her partner.

Lillith laughed, put her sword away, and her shield.

“Those guys were an embarrassment to any swordsman!” She laughed and hugged Andrew. “I wish I could say that was fun, but not even a challenge with the shield now.”

“You are amazing,” Jack said, which shocked everyone, a compliment from Jack? They all laughed

They left the town, went to the field they landed in, and flew off.

****

They flew, and still no luck. It was getting dark, and Lillith could tell Praisia was getting tired. "Ok, guys, let's land over there, no luck today....." Lillith said sadly. Then landed in an open field.

"Aww man." Andrew sighed. He knew how sad Lillith was.

"I'm just going to eat and go to bed so that we can get up early. I suggest you guys do the same," Lillith advised

They all agreed, they were all tired anyway. They walked towards the main road of the city to find a pub. The houses on the streets were scattered because this was a farming town, with huge fields of corn and wheat. Farms with pigs, cows. It took quite a while to find the street with the shops. They found a small inn. They would have to share rooms. Lillith and Andrew shared a room. That left Dom and Jack to share a room. The pub next door didn't have Cream soda. They were not happy, but they had big glasses of water. Lillith ate, then left. The others didn't go to bed quite that early.

****

Lillith lay in bed while the others were in the pub. She heard now seven towns were destroyed, burnt to the ground, and the ground was dead, the survivors rushing to the closest towns, no doubt hiding in the forest first, to get away from the wendigos. Thank the Gods they couldn't sniff them out, and their sight is really bad, because they are pretty much dead. It's like the head skinwalker puts a boundary around the town they are targeting, so he doesn't lose any of his precious army. The skinwalker must have a limit on how long it can hold up thebarrier, because it takes down towns one by one, every couple of days.

He is more interested in killing the land than killing humans, because more than half get away. But the poor ones that don't get away...... Lillith mumbled and hid her head in the pillow.

Praisia squeaked and lay on Lillith's head.

"Oh, come on, Lillith, don't blame yourself, now show me that face I fell in love with," whispered Andrew.

"But.....what about the poor dead...." Lillith pouted.

"Think about the living, I guess the head skinwalker thinks all humans will die when the planet dies, so he is being sloppy." Answered Andrew.

"Yes, we will fight for the living and the dead, we will make it up to them all," Lillith promised.

"Yes, we will, now close those eyes and sleep."

Praisia moved onto Lillith's shoulder. Lillith curled up. Praisia squeaked, then Bob squeaked back to his mom.

****

The next morning, they woke up at the crack of dawn, got dressed, and banged on their friends' door to get them up. They were out in minutes. They headed for the pub for breakfast, fed their partners plenty of food, and each had a glass of water. Merlo spilled his almostempty cup, and then they all lay on the table, squirming and sick.

"Merlo, what's wrong?" Jack raised up and was just about to freak out.

"Calm down, Jack, it's just because the planet is dying, I have the cure." Lillith reached into her bag and gave each partner a nut. They were back to normal in minutes.

"Phew." Jack picked up Merlo and hugged him. Merlo curled up in Jack's arms.

"Okay, let's get back on the road," Lillith said, but felt sad. Wait! She couldn't be sad, because today they could find one of their teammates! She had to keep up everyone's spirits.

Lillith giggled. "I have a feeling we are going to find them soon!"

"That's the spirit!" Dom said happily.

They flew off on a high note from the field next to the town, only two more to go!

****

No luck this morning yet again, Lillith ordered them to land. Break time for their partners, and lunch time for them, wonder what to eat?

Lillith wondered; she pet Praisia. "You are doing a good job of leading all the others."

Praisia purred, stood up, and stretched, then curled up and fell asleep almost instantly.

"Yes, you sleep, my angel, you are doing an amazing job," Lillith thought to herself.

They all sat around the corner table in the pub, which was strangely near where they landed. Why would a pub not be where the people go? Oh well, no big deal.

They ordered, and soon enough were eating, not to mention drinking their favourite drink, cream soda.

After they finished, Lillith wanted to have a serious chat.

"Okay, guys, I'm getting nervous that we haven't found a new member of our team. In a few days, we will have covered half of the continent. Any ideas?"

"Maybe they are dead?" Jack said nonchalantly, then laughed. They were trying so hard, and if it ended up that way, they would be in big trouble.

"That's not funny!" Lillith raised her voice and looked angry.

"Well.....?" he said, like what he said could be true.

Shut the hell up, Jack!" Dom said angrily.

"Heh," He huffed, no doubt to antagonise the rest again.

"Well, I just don't think fate would be that cruel, but shut up, Jack, you are such an ass." Lillith pondered. "We still have half a continent to find them, gotta be careful where the dead towns are, but I have a gut feeling they are okay."

They got ready and finished their drinks. The partners were waking up and stretching out their wings.

Off they were again, Lillith looking hard through the crystal, nothing. Lillith sighed and kept looking all over.

****

It was starting to turn dark outside. Lillith almost said, "Let's land," when she finally found a shining city!

"Found one! Let's go, follow us to land!" Lillith squealed in excitement, their mission was almost over.

They all landed softly in an open field and walked to the centre of the city, in search of a pub, that's where people go, because there's nothing else to do in a small town, but just in case, Lillith used the crystal to look at the people on the streets, nope. They found the pub and entered. It was smoky and dirty, but beggars can't be choosers. They sat at a side table. Lillith used the crystal, and she looked around. No one.

A girl with a long sword on her back came from the bathroom. Lillith held her breath and looked through. The blond girl with blue eyes was glowing! Lillith almost fainted, and she's trained with a sword! She was a beautiful girl. The sword was almost longer than she was tall. That is so cool. Lillith told them not to interfere; she would take care of this.

Lillith walked up to the girl and sat next to her. "Can I buy you a drink? I'd like to chat with you about something important."

"Oh, for the love of the Gods, I'm not interested in your religion, why bring that crap into the damn pub?"

"N-no, no, this isn't about religion, it's about the evil killing our world," Lillith said softly.

"I tried to join in the fight against them, but they wouldn't let me because I'm a woman! Isn't that stupid?"

"Yes, it is. I have a way you could actually be a part of a team that will destroy all the evil," Lillith said soothingly and calmly.

"What are you talking about? You are not making sense." The girl looks lost.

"If you want to understand, touch this crystal." Lillith puts the clear crystal on the bar.

"I guess it can't hurt." The girl reaches out and places her hand on the crystal. Her body jerks and she falls onto the bar, her eyes roll back into her head, and she shakes.

Lillith rubs the girl's back, "I know it's harsh, but now you know everything." Lillith soothed.

The girl sits up and looks excited.

"Are you ok?" Lillith asks nervously.

"I'm better than okay! I have a purpose! I am a legend! I will fight hard for this team. Can I meet them, the team?"

"Of course! Come." She smiled and brought her to the table where everyone was sitting.

"Hi, I'm Jade. I will be fighting with you guys. I use a longsword, and I'm a master of the art of the longsword. Oh my Gods, look at your partners! They are so cute, oh wow, dragons and a luck dragon. I wanna find mine as soon as possible." She squealed excitedly.

They all introduced themselves, and she made a great impression on everyone.

Ok, now that everyone knows each other, Jade and I are going to search for her partner.

****

They enter the forest, and nothing yet. They go deeper and deeper until they reach sand dunes, and her forehead starts to burn.

“We are getting close.” Lillith murmurs. Then she sees a seashell, and she runs to it, her head on fire. She opens it, and there is a beautiful orange egg with designs on the shell.

“Oh my God, is that my partner? So beautiful!” Jade carefully takes the egg and keeps it warm.

****

After hanging with her new group, Jade goes to her house, lies down, sends all her love to this little egg, places it on a pillow, and falls asleep. A few hours later, the egg starts shaking. Jade wakes up out of nowhere, then sees the egg rolling around. Then, POOF, it bursts out of the egg. The little orange dragon with red eyes, with three horns, locks eyes with Jade. It locks eyes with Jade, and it tells her her name is Magna.

“You are the most beautiful creature I have ever seen. May I pet you?”

Magna rubs against Jade’s hand, chirping and whistling.

Jade giggles. “Okay, I’m very tired, let’s go to bed, you can sleep on my shoulder if you want.”

Magna chirps and climbs up onto her shoulder and snuggles into her shirt.

They both fall asleep.

****

Lillith gets up early, wakes Andrew, gets dressed, then bangs on the doors again, then goes to the corner house, as the others start exiting their rooms, Lillith bangs on Jade's door. "Get up, sleepy head, we need to eat breakfast, then get on our way, we have to hurry, find the final team mate." Jade jumps out with her long sword on her back, with a little orange dragon on her shoulder.

"Oh, it's so cute!" Lillith never gets tired of looking at dragons.

"Her name is Magna," Jade announces.

"Ok, let's get breakfast, then we can be on our way."

Jade sits next to Lillith, "What are your names again?"

"I'm Dom, this is Umi," He points to the blue dragon on his shoulder.

"I'm Jack, this is Merlo," Jack points to the luck dragon..

Lillith points to her partner. "This is Praisia."

"Oh, I'm Andrew, and this is Bob."

"Nice to meet you all, I'm Jade."

They all walk to the pub, order breakfast, and eat a nice meal, sharing it with their partners.

"Is everyone done with their breakfast? If so, let's go! We're burning daylight!" Lillith pushed.

They walk to the field outside the town.

Magna grew big as soon as all the other dragons grew, smart dragon! Lillith liked that. Lillith was impressed.

They all flew up into the air from the field they landed in. Lillith took out the pink crystal and started looking all around.

****

Lillith commanded everyone to land. It was lunchtime, and their partners needed to rest. Even if Lillith wanted to be selfish and keep going, she had to take into consideration all the hard work of Praisia and the others. Lillith was getting very frustrated, as if it were looking for a damn needle in a haystack. There are so many towns to check. She sighed and went to the main road to look for a pub. Everyone followed her. She was going to lose it. They do the same damn thing over and over and over again. But she didn't have a choice; this was her mission, no matter how annoying it was. They sat down and ate. She didn't even care what the others were having; she just wanted to get back into the sky. Lillith listened to Praisia breathing, and she realised how exhausted Praisia was. Her heart broke; they were trying their hardest to help. But she had to get them to fly to find the last member. Lillith realised she was slipping into a funk; she had to get out of this. She lay her head on the table and fell asleep. Maybe sleep will help?

****

"Lillith, Lillith, look to the east, near the dangerous part of the continent that is being taken over. Hurry, before it's too late! WAKE UP!!" Lillith jumped out of the chair.

"I know where the last person is, I got help from someone! We have to hurry! It's in the danger zone! We don't have much time!!! It's east, by the parts that are being taken over."

"Um, you are hearing voices in your head......coookuuuuu," Jack replied.

"I believe her," Andrew said, standing up.

"Me too." Dom stood up too.

"I'm in!" Said Jade. "I believe in Lillith!"

"Okay, let's go," Jack said, rolling his eyes at the whole idea.

They asked their partners if they were ready to fly yet. All of them squealed happily.

"That must be a yes." Lillith laughed.

"Ok, so there's going to be dark clouds and smoke, ruined towns, maybe even bodies. Are you guys ready for that?" Asked Dom.

“We can take anything to get our mission finished, then we can save the world!” Jade said proudly.

“Okay, let’s get out of here,” cheered Lillith excitedly.

They flew into the sky from the empty field. “Okay, we are turning to the eastern part, where the evil is,” announced Lillith.

They flew until they started seeing dark clouds and smoke rising from the ground. They saw the dead ground, which was very upsetting, but they couldn’t stop it. Lillith took out the crystal, there!, right there! Oh God, it’s only got one town between it and the dead ground. “Guys, land on the northern field, to stay away from the death lands.

They landed softly on the northern field and hurried to the town.

No one was on the street. All shades were closed. The only thing open was the pub.

“Well, I understand why it’s still open, I mean, where else are you going to go to drink your woes away?” Jack pointed out.

“Well, they are in this town, so if they are not in here, we might have to go door to door,” explained Lillith.

“Why are they all still here? Why don’t they run away?” Wondered Jade.

“Where are they to run?” Jack replied.

“To other towns, to a faraway town!”

“There’s no room; they would become a burden. People are pretty stubborn around here.”

“Ugh, I don’t like that.”

“Oh well, what matters to you doesn’t to them.” Jack laughed, then followed Lillith to the pub.

The place was packed.

Lillith took out the crystal and looked around. Too many people were in the way. She scooted around, then there she was. She was sitting in the corner, not talking to anyone.

“Okay, I got this,” Lillith murmured. She walked to the table, Excuse me, may I sit here?” Lillith asked.

“Sure.....”

“May I ask you a few questions?” Lillith questioned.

“Sure.....”

“Do you have family close by?”

“They are all dead.....” Answered the girl.

“How about some close friends?”

“Nope, kinda a loner.”

Do you want to leave here?”

“Yes, I want to get far away.”

“That’s perfect! We need you to help us fulfil our destiny.”

“Destiny? Now that sounds interesting.” The young woman turned her full attention to Lillith.

Lillith took the crystal out and placed it on the table.

“Touch that and all will be known.” Lillith directed.

The young lady placed her hand on it. The same thing happened all other times. She shook, her eyes rolled up in her head, and then she flopped down on the table.

The young woman raised her head and looked Lillith in the eyes.

“I’m scared, so scared!” Cried the young lady.

What could she say to that? It was scary.

“Come with me, we need to find your partner as soon as possible.” She held her hand and led her out. She walked around the stones that were prominent in the area, and her forehead started to burn. "Oh, thank the Gods, it's around here." She looked, and then it burned so much. Lillith lifted a stone, and there it was: a beautiful Black egg with swirls. "Here, this is your partner. You give it lots of love, and it will hatch right away."

"Can we please leave this place.....I'm from two towns down, those bastards murdered everyone I knew and cared for, I'm going to end them."

"Yes, let's leave right away. Just need to get the rest of the group." They were exiting the pub, and Lillith waved them over. "Let's get out of here!"

Lillith let the girl, ugh, she didn't ask her name, dammit. "Next stop, my home."

"Is your home safe?" Asked the girl.

"It's very far away from your town, so I'm hoping it is. It's on the other side of the continent."

"Ok, good."

"So, what is your name? Mine is Lillith."

"It's Selene."

"That is a beautiful name." Lillith complimented.

"Thanks."

"Do you know how to fight with weapons? It's a strange question, but if not, we will have to start from scratch."

"I was taught to fight with a scythe, with a sharp knife at the other end of the pole, oh and the scythes blade was sharp on both sides so that I could do more damage, I also know how to block attacks with the pole, best in my class, my town was a farming town, but had a lot of bad people that you had to learn to protect yourself, or you were just asking to be killed."

"That's scary, but very interesting, going to have to have you go to the blacksmith and make a strong one for you, top notch."

"Oh no, that's ok, I can use a regular one."

"After hearing all about that one, I wanna see it!"

"Yea, it's pretty badass." Selene giggled.

"How is your egg?" asked Lillith.

"It's nice and warm."

"Good, I'm glad there is another girl in our team, I was feeling like there were too many damn men!" Said Lillith.

“Ha, yea, I noticed that. How sad.”

“Very!” Lillith laughed. “Oh, and I’m very sorry you lost your whole family....”

“It’s ok, they all hated me, they treated me like I was the black sheep of the family, I mean I’m sad they suffered and died, but my life will not be changed at all because they were never a part of my life, left out, ya know?”

“Oh no, that is terrible, they should have treated you better.”

“Shoulda, coulda, woulda, but didn’t, their loss, not mine.”

“So, do you like the color of your egg?”

“It’s my favourite color,” Selene replied.

“Well, that is the color of whatever hatches outta that egg.”

“Oh, good, and I will be able to ride it right away? That is what the crystal pointed out.” Selene asked.

“Yup, if you fill it with lots of love, it will hatch into an adult, but a tiny version that grows big by choice.”

“I never knew dragons existed, thought it was just a made-up fairy tale.” Selene thought out loud.

"Yea, I think all humans thought that, but these eggs, whatever they hatch into, are different than regular dragons, because I don't think dragons can shrink at will, but that could be normal."

"Interesting." Selene wanted her egg to hatch, but it might be too dangerous up here, so she hoped it didn't.

"Well, enjoy the ride to safety, and be prepared to spar with us.

"I will."

Neither of them said another word until they were at Lillith's home.

Chapter 5 – Wings over a Broken World

CHAPTER 5

WINGS OVER A BROKEN WORLD

They finally reached Lillith's hometown, and they landed right beside her home. She climbed down and helped Selene off Praisia. Praisia then shrank and flew to Lillith's shoulder. Okay, guys, wait here. I just need to tell my parents where I have been. She opened the front door. "Mom, Dad, I'm home!"

Lillith's mother ran to Lillith and started crying.

"Mom, it's okay, I had to gather my team, because we are going to save the world from evil," Lillith said.

"What? We thought you ran away from home with Andrew!" She sobbed.

"No, Mom, it is much more complicated and important than that." Lillith laughed.

The door slammed shut. Lillith's father had entered.

"Why are there a group of people in my yard, and where have you been?"

Those are the warriors that are going to help me get rid of the evil in the land," Lillith explained.

"What's this nonsense you are jabbering about?" Asked her father.

I had to go on an adventure to find them. It was so hard, but yes, we are going to rid the world of evil. See this mark on my forehead? It means I'm special. I don't know why, but it stands for something. I gathered the four warriors who will help me get rid of the evil. It's the only way to get rid of the evil."

"I knew there was something special about you," the father admitted.

"What? You actually believe this bullshit?" The mom shouted at him.

"It makes sense, hun, look at her, does she look human? She's far too pretty to be a mere human, not to mention her strengths and abilities," he questioned the mom.

"But she's our little girl, she can't be in this pretend war on evil...." She covered her face.

"I just came to tell you guys I'm okay and have to go back out to do other things to prepare for the big fight," Lillith explained.

"Wait, we need to tell you something."

"Hmmm?"

“Your mother and I couldn’t have children because of female problems, but then we found you on our doorstep, and thought you were a gift from the Gods. We still feel that way.”

“Oh my Gods! You will always be my parents.” Lillith ran to them and hugged them.

“Go on your adventure, but please don’t forget us.” The father said sadly

“I’m never going to forget you, I will come to see you as often as possible, but now I have to go see the Elder witch. Love you both.” She hugged and kissed them both on the cheek, then ran out the door.

“Okay, guys, let’s go see the Elder witch.” Lillith led the way. They walked while everyone stared at them.

****

Lillith knocked on the Elder witch’s door. The Elder witch opened the door and clapped her hands together.

“You did it!” She put her hands on everyone’s face except Jack’s because he refused to let her touch him.

“Gross!” Jack spat out and was ignored.

“Now what?” Lillith asked.

“You must gather up your true weapons, but this task is much easier. You must go to the island west of here, then the witch will use your souls to create your perfect weapon, one that will never break, can never be used against you, and no one can control them either. They are in-tuned to your very being.”

Could you point it out on a map, Elder witch?” Lillith asked.

“I can’t, it’s not on any maps, just keep going west after the beach on the west side. There will be a rock on the beach that has an arrow pointing to it, just keep going until you see land, he will be expecting you.”

“My egg is hatching!” They all turned to look at Selene’s hands. The egg rolled, and then, poof, the dragon popped out. It was a beautiful black dragon with orange eyes and one horn. It stared into Selene’s eyes, “His name is Otto!” He crawled up to Selene’s shoulder and cuddled her neck.

Selene squealed.

“Okay, are you sure there is an arrow pointing in that direction?” Lillith needed reassurance.

“Yes, we witches put it there so we could find it by boat. And be glad you have partners to ride, boats were hellish to cross the ocean.”

“Ok, this task seems easy enough. We will be back as soon as possible.” They stepped outside. Their partners knew they were needed, so they grew. they all flew to the western beach.

****

They reached the beach.

"Now to find the arrow." Sighed Lillith.

Selene pointed out a huge boulder in the shape of an arrow.

"Okay, let's follow this arrow to the island," Lillith said.

The partners all looked that way and got ready, as they were mounted.

"Let's go in a line, one behind the other, I will lead, okay, so we won't get lost?"

"Okay, they all agreed except for Jack, who was annoyed he wasn't the leader of this quest. He rolled his eyes at them, "Whatever."

They took off, all followed one by one.

After an hour of flying, they wondered if that was the wrong rock. But just then, an island came into view.

"I see it, guys, almost there!" Lillith yelled back at the rest.

****

They landed on the rocky beach, one at a time. All the partners shrank and flew to their respective partners.

Lillith looked for a house, but all they found was a cave. Lillith started into the cave and called out. "Witch?" She heard a humming sound from inside. She went deeper, but everyone was following her.

She called out again for the witch, and she then found him sleeping in a comfy chair, resting his head on his arm on a metal desk. Lillith approached the sleeping witch, then touched his head. He jumped up.

'Woah! What are you youngins doing on my island, get off!" Lillith bent down to his height and told him the Elder witch sent them. The witch wiped his glasses, then looked at her forehead.

"Oh dear Gods, it's really happening!" He stood up and stretched. "Okay, so I need to make all of you your essence weapons, except for that boy." He pointed out Andrew. Andrew laughed.

"Ok, first up is air, who is the element air?"

They had no clue what he was talking about.

"Oh, come on, she didn't tell you, the four people you collected represent an element. Now let me look and smell you lot," He walked to Jade, he smelt her, You are fire, he smelled Dom, you are water, He smelled Selene, "Earth for sure, so that leaves you boy, let me smell you to be certain."

"Ew, no." Jack flinched.

The witch flew at Jack too fast to dodge. "Yup, air, follow me please."

Jack looked violated, but followed anyway. They waited; it took about twenty minutes, and Jack came out.

"Earth, next please."

Selene went into the room. Twenty minutes later, Selene came out.

"Water, next please."

Dom went into the room, and twenty minutes later, Dom came out.

"Fire, next please."

Jade went in, and twenty minutes later, Jade came out.

"Mystic one, next please."

"Does he mean me?" Asked Lillith.

"Yes, my dear, you are the Mystic one, please enter."

"You didn't smell me, how do you know?"

"That marking on your forehead, you are special."

Lillith went into the room, and it was dark, but you could still see enough.

“Please enter this case.”

It was like a coffin, but it was standing up and made of some type of rock.

“Enter please.”

Lillith entered, and he shut the door behind her. She heard a rumbling, more like rolling, then warmth, then dizziness, and a loud clap. And the witch opened the door.

“Not so bad, hmm?”

“No, it wasn’t bad at all. She smiled at the witch. Then left the room, looking like this really was easy.

The witch left the room.

“Now, to let them all settle, it will take a few hours. So join me at my house for a fine meal, roast beef with carrots and potatoes.” All made by magic, I never touched anything. So don’t be scared to eat.” He looked at Jack and scoffed. “Please indulge yourself.”

They all followed him out of the tunnel. He waved his hand, and a house that was not there a moment ago was there now. They all entered, and inside, there was a big table with a roast beef, cut up, and carrots on the side with potatoes. They all had some of the yummy food, including Jack.

"Jack, you said you are a witch, how come you don't know the stuff they do?" Asked Jade.

"Oh, well, I'm a wandering witch, we don't stay around and learn from the elders and the witches like this guy," Jack explained.

"Oh? So you have never been strictly taught to be a witch?" Asked Dom.

"I was taught, but most witches do this when they settle down and become witches like the Elder witch," Jack explained more.

"Ah, okay, gotcha." Jade was satisfied with Jack's answers.

They all played card games until the witch announced that the weapons were done.

The witch handed Jack a long staff, silver in color, with a fancy design at the top, that looped around nothing, and a very sharp knife at the bottom of the staff. It was light but sturdy, and Jack was very happy with it. The witch handed Jade another long sword, but this one was a black blade with a reddish hue that almost looked like fire on it. And sharp on both sides of the sword. Jade liked the weight and balance of it. She was very excited to use it. The witch then handed Dom a sword, a very fancy sword, very light and easy to swing. If you dropped a hair, it would split the hair. It was silver with blue tones. The witch handed Selene a scythe, a black pole and a black scythe blade, sharp on both sides, then a sharp fancy blade on the bottom of the staff. The witch then handed Lillith a long sceptre, with a hole in the top, which had fancy designs from top to bottom.

"What's this?" Lillith asked the witch.

"It's the most powerful sceptre in the world. You are going to need that for the fight."

"But I thought I would get a sword, I have no idea how to use magic."

"You will, once you get your element, which is Mystic."

Lillith blinked and didn't know what to say.

"All of you have weapons, but once you get your elements, you will have to use them to slice and thrash. Your elements will give you power, so don't think you will get destroyed by the army of wendigos or even the skinwalkers. The only thing to worry about is the dark crystal that the head skinwalker uses. That is where The Mystic comes in. She must combine all the elements together to create the Light crystal and battle the dark crystal. But you have to find the elements first, which will not be easy. I'm sure Elder Witch will have something to make it easier. Oh, and when you get your elements, your partners gain power too. But you will see that."

All were very stunned and not sure what to say.

"You may leave now."

They all left quietly, their partners jumped down and enlarged.

"Oh, and follow the arrow to get home."

Jade found the stone arrow, and they all left the way they came, one after the other.

****

When they reached Lillith's hometown, they went to the pub to talk.

"So, we all have elements we need to find, and they are crystals, too. What are your thoughts, oh, and our partners will be joining the fight, so we have to train when we get all of our elements," Lillith commented.

"My question is, why is he?" Jack points to Andrew. "Still joining us on this?"

"He gives me strength, the Elder witch even said he SHOULD come along." Lillith pointed out and held Andrew's hand.

Andrew smiled. "Yea, I'm her support."

"My lucky charm." Lillith joked and kissed him softly.

"But Jack has a point; he could get killed," said Selene.

"He has done nothing but make this team stronger. Please don't question me unless it is a serious issue," Lillith said with a little anger in her voice.

"Okay, but if he dies, it's on you," Jack said snidely.

"I won't let that happen, neither will Bob or Praisia."

Jade cleared her throat. "Maybe he is in this group for more than just support, I mean if the Elder witch says he should be with her, then who knows, I mean this is magic!"

"Thank you, Jade." Lillith smiled. "So are you guys happy with your elements?"

"Yes, I am!" said Selene.

"Me too, I mean it should have been obvious, my hair is blue," laughed Dom.

"Yea, how is that natural?" Asked Jack.

"Dunno, but it runs in my family, my dad had blue hair, it's like one person every generation gets this hair," Dom explained.

"Well, I think it's very nice," Selene commented.

"Jack, how do you feel about your element?" Asked Jade.

"I like it, wind can destroy buildings and so on." He laughed.

"Yea, I like mine too." Jade smiled.

"Hun, what do you feel about yours, Lillith?" Asked Andrew.

“I don’t even know what mystic is? And I mean, I have fought with staff and so on, it’s just I figured I’d have a sword too.”

“hun, yours isn’t like Jack’s, yours is a sceptre, not to mention the most powerful one.” Andrew looked at it on her back. “It’s beautiful too. Suits you. Actually, all of your weapons are beautiful and suit you guys.”

They all smile at Andrew.

“You make it really hard to hate you, you know that, right?” Jack grumbled.

“Thanks, Jack, that’s quite a compliment.” Lillith grinned.

“No problem, just saying the obvious,” Jack replied.

“Okay, guys, it’s getting late, we should go to bed, we had a long day, and we need to visit the Elder witch to start our journey to find our elements. Well, I will be getting you all your own rooms. If there’s not enough room, you guys can talk it out.”

“Thanks, Lillith.” Selene smiled at her.

****

Lillith lay on the bed, crumpled up in the covers after taking off her armour and weapons.

Andrew sat next to Lillith and rubbed her arm. “Are you okay, hun?”

“I’m just scared, what does mystic mean, what is expected of me, I have a feeling I have to do something bigger than I can comprehend, something to do with all the elements....I know I’m not human, but am I strong enough?”

“My sweet Lillith, I have never met anyone stronger than you; you will do good, and I will always be here to help you.” Andrew leaned over and kissed her cheek.

“I could never do this without you, you know that?” Lillith said quietly.

“Yes, you could, you are a lot stronger than you let on. But I will always be with you, my future wife.”

Lillith smiled, turned around, and hugged her beloved Andrew. Praisia and Bob flew to the towels and curled up together. They felt they should let them sleep without bothering them.

“Let’s cuddle until we fall asleep,” Lillith suggested.

“That’s exactly what we are going to do.” Andrew laughed and lay down beside Lillith, then held her head to his heart. “Do you hear my heart beating?”

“Yes, I do.” Lillith giggled.

“Every beat of my heart is for you, you are my soulmate, and I promise to try and always keep you happy.”

Lillith began to cry. “Do I really deserve someone so great?” She wiped her eyes and hugged Andrew close.

Andrew held her, too. “I think we should sleep now, my dear, you have sweet dreams, I love you.”

“I love you too, with all my heart.”

They both slowly fell asleep.

****

They all met in the pub for breakfast.

“I dunno about you guys, but I’m starving!” Lillith announced.

“Me too,” said Dom and Selene, and Jade at the same time.

“I could eat.” Murmured Jack.

Andrew was busy playing with Bob. “Oh, I’m starving, so is Bob.” Andrew agreed.

They ordered a big breakfast with a glass of cream soda, then left for the Elder witch’s place.

When they got there, they knocked on her door.

The Elder witch opened the door with a big smile on her face.

“We got our weapons, Elder witch,” Lillith announced.

“Good good!” She laughed.

“We found out we need to find our elements. How do we accomplish that task?” Asked Lillith politely.

“Oh that big mouthed old fool, he probably made you guys all nervous, I have a magical compass, it shows you the elements when you are near them, now i have to tell you, your elements are not crystals until they combine to your essences, but there is so much you guys have to do before the fight, you must learn your powers you will gain, then after you defeat the army, you must combine the crystals to create the light crystal to defeat the corrupted dark crystal. that will all be on you guys, then Lillith will have to fight the head skinwalker. Then there’s even more to do, but I won’t get into that until the time comes. But your journey is not close to being over.

“When we join the elements together, will we still have the powers?” Asked Selene.

“Of course you will, once bonded, nothing can break that, even if the crystal is not with you anymore.”

How do we combine them?” Asked Lillith,

“I will teach you when you get back from this journey.”

“How many journeys are there?” Jack said, frustrated.

“Plenty, my impatient one,” the Elder witch pointed out.

“But Elder witch, people and the world are dying, we don’t have time for all these journeys,” Lillith explained.

“Yes, I understand that, but there have to be sacrifices along the way. I know it seems cruel, but I’m sorry, there is no other way. If I knew a faster way, I’d tell you.”

They all sighed, even the partners.

“Okay, now that you understand, let me get the compass.” The Elder witch went into the back and took out a gold compass. “Ok, come here, dear.” The Elder witch asked Lillith. “Now look at this, see how the face is all black, when you get to where the element is around you, it will glow in that color that represents the element. Red for fire, blue for water, green for earth, yellow for air, and lilac for mystic. Land, then the person gets down there and concentrates on the element. Then they should receive the element, drawn from the ground. Then it will become a crystal. You have to call it out in your mind, draw it close to you, and it will become one with you. When you all get this done, come back here. I have to teach you to use them and fight with them. After the army is all gone, join them together. We are almost done with part one, you don’t even know what great things are going to happen because of all of you, yes, including you, Andrew.” She winked at Andrew, and Andrew smiled back at the Elder witch.

“Okay, thank you, we are off again, will they be in non-populated parts of the world, or in towns?” Lillith hoped she could help her with this.

"Non-populated places, but to help you, they are all on this continent, so no worries, just stick to forests and deserts and swamps and so on."

"Thank you so much! This will make it much easier to find. See you soon." Lillith said, and then they all left.

****

"Okay, there are so many places to go, maybe we should start in the forest, it's closest and we might find multiple elements." Lillith thought out loud.

They all agreed on this. "So we are onto the Ebony forest, it's the biggest, wildest, and purest forest." Lillith directed.

"Sounds good." Dom agreed.

"I dunno about you guys, but I'm getting nervous," admitted Jack.

"I agree." They all said at once.

"But we are gonna do what we are destined to do, because we are all a great team." She smiled, "Okay, let's go!" Praisia jumped off Lillith's shoulder and grew. They all got on their partners and flew towards the Ebony forest. Lillith took out the compass and scanned the ground. They were halfway through the forest, and still nothing. Lillith bit her lip. She was stressed out because so many towns had been attacked, and the land was in ruins. She wasn't sure how many towns now, but a good amount no doubt. Then green filled the black glass on the compass.

“Guys, it looks like Earth is first! Let’s fly down. They had to be very careful because branches could hurt their partners.

Selene gulped. She was so nervous being first; not seeing others do it made it much scarier.

They all got off their partners and looked at the compass.

“Looks like the centre is right there by the stump, okay.” Lillith knew how scared Selene was, so she took her to the side. “You can do this, Selene, you were born to do this, trust me, trust your partner, and trust yourself. Now go sit on the stump, and focus on the element, it will come, I believe in you, we all do.”

Selene felt so nervous that she’d not be able to do it, that she would fail. Selene took a deep breath, then walked to the stump. She had to do this for everyone. She sat down, closed her eyes, and pictured a green glowing light. Then, out of nowhere, a gust of wind rose from the ground, swallowing Jade. But Selene was now in a trance, and the wind turned green and enveloped her. It felt warm and calming. It felt so good. She felt the energy was filling her up. Then Selen’s weapon glowed and formed a green crystal on top of her scythe. The wind died down, and Selene opened her eyes. She grabbed her floating weapon and placed it back onto her shoulder. She looked at her partner, and it now had a green crystal on its forehead. Jade smiled at her partner and then at everyone. “I did it!”

Lillith ran to Selene and hugged her. “You did amazing! We are all proud of you, you were so brave!” They were all excited now.

“Okay, it’s getting late, we need to eat and relax, we found one already, this is just amazing!” Lillith exclaimed.

They all hopped on their partners and flew to the nearest town. They didn’t even care about the name of it. They were hungry and tired. They walked around the town and found the pub. The inn was right across the street. The houses in this town were all made of wood, but they were very nice. This must be a nice town to live in. They all thought.

They entered the pub and sat at the corner table. They ordered their food and cream sodas.

“So how does it feel?” Asked Jade.

“I feel like I’m buzzing with energy, it’s boosted my confidence, that’s for sure.” Selene giggled.

“I think we should try the desert tomorrow, I have an inkling that one will be there,” announced Lillith.

“Sounds like a plan.” Dom agreed, and so did everyone else.

They ate, then decided to go to bed early, so they got up early.

****

They all woke up and met everyone outside the inn.

"So the desert is pretty far away, we will have to take a break for our partners, okay?" Lillith explained.

"Yea, at the edge of the desert should be a good place to rest." Said Selene.

"Yea, that should be perfect, but first, breakfast." Lillith rubbed her stomach. "I need food."

"Ok, let's go to the pub," Andrew said.

They all had a big breakfast, drank their favourite drink, cream soda, went to the bathroom, and then they left.

It took a few hours to get to the desert. It was near the towns that were being attacked, the great Silia desert. They all just sat and relaxed, asking Selene many questions, like how she did it, and told her she almost glowed the color of the crystal. They were excited to get their own powers. Lillith was quiet. She cuddled against Andrew. She was very nervous. She was the leader, but why? She didn't feel she deserved it. She whispered to Andrew.

"Hun, I don't feel worthy as a leader." She admitted.

"You are so worthy, you just don't see how special you are, trust me, everyone believes in you."

Lillith sighed and rested her head on Andrew's shoulder. "I just hope I don't let the world down."

"Think positively and you will go further, negativity only stunts you, only think of the positives, you are amazing, and strong, you are the perfect person for this." Andrew kissed her forehead, then rested his head on hers.

****

After the partners had rested at the edge of the desert, they decided to get ready for the challenge of finding the next element. They flew around abandoned old towns. Lillith kept a watchful eye on the compass. Nothing yet. She was getting worried because they were running out of desert, but at the very edge, a yellow light filled the compass.

"Ok, guys, let's go down! It's Jack's turn!" Lillith laughed.

Jack almost choked on his own spit; he was next, but he had doubts about himself, as anyone would in this situation. But he was going to at least try.

They all landed.

"Okay, the centre of the light is that small boulder, go sit on it, and let it in, Jack," Lillith said.

"It's not hard to do, Jack," Selene told Jack.

"Okay," He wandered over to the rock and sat. Merlo rubbed against Jack's cheek. "Thanks, buddy, I will do my best." Jack closed his eyes and concentrated on energy, he could feel something, something warm, he concentrated on that warmth, it felt like air was whipping around inside of him, then wind swallowed him, a yellow wind, it filled him up, his staff lifted up and floated, a beautiful yellow stone filled the empty space in his staff. Then his partner's forehead had a new yellow gem on it. Jack grabbed his Staph and placed it back onto his back, then looked at his hand; it felt just like Selene had said, buzzing with energy. This wasn't hard, this was meant to be! Jack walked to the group and asked if they could go to a pub, eat, and rest. He felt dizzy and needed to rest.

"To be honest," Selene stated. "I felt the same way; it really takes a lot out of you."

"Oh, okay, I guess we can't rush; this is a task that needs to be taken slowly." Lillith smiled. "Let's go to DaVille, it's kinda close." Offered Lillith.

"Sounds good to me, let's go," said Jack.

They all flew to the town and went to the pub to relax. They ate their dinner and drank their cream soda.

"This is really getting exciting!" Said Jade. "I mean, you two are like glowing, I can feel your energies radiating off of you. "I want my powers too!" She giggled.

"It definitely is different," Admitted Selene

"I know it's strong, but I'm not about to use it without being taught how to, I'm a witch, and I know better," said Jack.

Selene scrunched her face.

"Bad things can happen using them without knowing how to use them, then?" Jade asked.

"Yes, terrible things can happen, not only to you but to the people around you, so don't even try," Jack exclaimed

"I wasn't going to, I was just wondering. Sheesh, do you take me for a fool!" Selene snapped at Jack.

Lillith was deep in thought of where to go next, oh, the mountains!

"Hey guys, we are going to go to the mountains tomorrow, on the old lava flow. I have a good feeling about this, is that ok with you guys?"

"Hell yea." They all cried. Lillith's hunches have paid off so far, so why not the mountains?

Chapter 6 – The First Warrior

CHAPTER 6

THE FIRST WARRIOR

The next day, they had breakfast, then left for the mountains. while they were flying there. Lillith used the compass, but nothing. It took a while, but they made it to the mountains. She looked at the old lava flow of the mountains, but saw nothing. She had Praisia land to give them a break.

"Praisia, can you fly up the mountain?" Lillith asked.

Prsaisia snorted and nodded.

"Okay, in half an hour we will all fly up the mountains, hopefully we will find the magic."

Jack offered everyone some beef jerky and water. They all had some, including the partners.

"Thanks, Jack." Jade smiled at him.

I tend to always carry food and water just in case." Jack said happily.

They waited. Then, Praisia snorted again, and Lillith knew they were ready.

They waited. Then, Praisia snorted again, and Lillith knew they were ready.

"Okay, let's go, guys!" Lillith said excitedly.

They all got on their partners and they started the high flight up the mountains. Lillith used the compass the entire time, then saw nothing, but then, after getting close to the top, there was a plateau, and it lit up red on the compass.

"We found the spot where Jade gets her powers." Lillith laughed and cheered. It was red, so that is not water or her own.

They all landed, and Lillith told Jade to sit on the lump in the middle of the compass.

Jade agreed. The wind blew, and they all realised how cold it was. They had to take the cold; they didn't have a choice.

Jade and her partner sat and closed their eyes. She pictured a flame in her chest, and it started to warm up. She concentrated hard. A red wind enveloped her, and she could feel the energy fill her. It was so relaxing. Then, her long sword floated off her back, and a beautiful red crystal embedded itself on the hilt of the sword. She grabbed her sword. Next, her partner began to glow, and a red crystal appeared on its forehead. Jade grabbed her sword and opened her eyes, giggling. She petted her partner and ran towards the freezing group.

"I feel so strong! But I'm very tired." Jade yawned.

"You did great!" Lillith cheered. "Okay, let's go to DaVille. I need to get a map and study, because I have no clue where else to go for Dom's or my own." Lillith pouted. "I need to get a map, it might help me."

They all agreed and off they went. When they got there, she told everyone to go to the pub, and she would look for a map, then join them.

The store on the corner has many maps, best to go there," informed Jack.

"Thanks, Jack." Lillith smiled, "Go with them, okay, Andrew?" Lillith told him.

"Okay, I will see you soon." Andrew kissed her on the forehead, then left with the group.

Lillith went to the store and looked through the maps; there must have been fifty different maps, but then she found a map that showed strange places. She opened it, and it was dotted with many different places.

"Perfect!" She folded it up and paid for it. Lillith then entered the pub.

"Took you long enough!" Dom jokes. "We waited for you to order." Lillith laughed and sat down.

"I'm craving beef, I think I might get a steak, medium rare!" Lillith was starving.

They all ordered and got a large cream soda.

After eating, Lillith yawned. “I don’t know how long it’s going to take me to find the right places where Dom and my magic are so that we might be staying here for a while, but I will do my best and follow my gut.”

“No worries, we will wait for however long it takes,” Jade said happily.

Lillith yawned again.

“Okay, that’s it, you are going to go to bed right now. We will keep Andrew busy. You go relax and sleep. You are working hard to lead us, and we know this. Please go to bed,” begged Selene.

“Okay, okay, Andrew, be good!” She laughed, then left the pub for the inn. She put everything on the table and flopped on the bed, cuddling Praisia. “Praisia, I hope we are doing good, I don’t want the world to die...”

Praisia licked her cheek and cuddled her to soothe Lillith. Lillith fell into a deep sleep.

****

Lillith woke up and stretched. “Ugh, what a wonderful sleep!” Praisia stretched, then jumped onto Lillith’s shoulder. Lillith noticed Andrew was sleeping on the comfy couch. Lillith was very quiet; she decided to let him sleep in. She took a shower and got dressed, then took out the map. There were so many different places, it took her a while, but in the middle of the map was a place called Dragon's Lagoon. Her heart skipped when she read it. It was in the middle of the Devil's forest, so it wasn't visited much because people thought the devil was in the forest, ready to eat them up. "Okay, one place, let's see if I can find the last place. She looked all over, then noticed a place on the corner of the map, called Crystal Henge. "That has got to be it, her heart felt it was the right place.

Lillith folded up the map, then left to go to the pub. She didn't want to wake Andrew up, so she left. She entered the pub and ordered two cream sodas, one for her, the other for Praisia. She ordered some breakfast, with extra sausage, for Praisia. She petted and played with Praisia while everyone slept in, which they deserved to do. She was having a good, peaceful time. Prasia loved the cream soda and ate two sausages and some eggs. Lillith ate the rest.

Lillith took out the map and showed Praisia the two spots. Dragons Lagoon was closer, so they would go there first.

Andrew looked into the pub. "There you are." He smiled.

"Yup, I'm here, go take a shower and then join us, sweetheart," Lillith ordered.

"Okay, I will be back soon."

Everyone was in the pub by noon.

"Okay, I have two spots I would like to check out, so finish up, go to the bathroom, and be ready," Lillith ordered.

"Anything you say, Leader!" Dom smiled.

****

They all took off following Lillith and Praisia. Lillith checked the compass the whole way there, just in case. They reached the Lagoon, and it glowed blue.

"Guys! I guessed right! Let's go down now!" Lillith yelled so they all could hear her.

They all flew down and landed softly. The lagoon's water was such a beautiful blue, it looked like it didn't belong in this forest. There were stepping stones to the middle, and a flat rock was in the middle.

"Go for it, Dom!" Lillith cheered him on.

Dom looked nervous, but he took a deep breath and started towards the middle. When he got there, he relaxed and sat down. He pictured a blue light and cool water flowing through his chest. Just then, a blue wind came up from the water and surrounded him. The light in his mind turned warm, and he could feel the magic enter him. His sword left his back, and a beautiful blue crystal formed on the blade near the tip. Then, a blue crystal formed on his partner's forehead. Dom opened his eyes and grabbed his sword, and laughed as the blue wind slowly went away.

"I did it, guys!" Dom laughed and cuddled his partner, Umi.

They all cheered him on and were glad their journey was almost over. Just Lillith's power now, but what does Mystic mean?

The woods creaked and cracked, then out of nowhere, five wendigos jumped out and hissed at them. Dom screamed to warn them, Lillith could smell them, and turned around to see the five wendigos.

Lillith grabbed her sword and shield. "Guys, we have to destroy them! We can do it." Lillith said with confidence, they gathered and took the right stance. As the Wendigos attacked, they fought back hard. They couldn't use their magic, but their fighting skills were good enough, and their weapons were strong. The first wendigo jumped at Lillith, and Andrew blocked her and sliced one of its arms off. It acted like nothing happened; the other four attacked the others. Dom sliced open the chest of his attacker. It didn't even notice and continued to attack.

"What the hell, it's like they are dead," Dom yelled.

"Yea, I think they are Dom! Guys, cut off their head, or legs and arms if you can, that's what will make them easier to kill." Lillith yelled.

Jade licked her lips and thought, "Come and get me, you bastard!" One jumped at Jade. Selene laughed and used her scythe, chopping off its head as if it were a hot knife through butter. It fell and stopped moving.

Lillith pushed Andrew behind her. "You are not meant for this fight! I love you, but I will be fine," She dodged the first wendigo and sliced it in half. It lay there and used its arms to crawl towards her. She jumped up and landed on its head. It did not like this; it screamed and struggled, then with one swipe, she chopped its head off, and she landed beside it.

Jade took her long sword and sliced the wendigo into pieces as fast as any other members of the group. She laughed, then noticed it was still alive. She stuck the sword through its head and flung the head into the dark Devil's forest.

Dom glared at the wendigo that kept swiping and attacking him. He was getting annoyed, so he sliced both arms off with two swipes of his sword. It growled at Dom, but Dom kicked it and it fell onto its back, then Dom swiftly sliced its head off.

Jack stabbed the wendigo that was attacking him. Jack then realised how sharp his staph was then sliced through its chest. Then used the sharp end to slice through its left arm, and it growled. "HUNGRY" and jumped on him. Jack was now on the ground with this freak on him; its smell was overwhelming. He blocked every time it tried to slash him with its claws and bite with its nasty, rotted teeth. He then kicked it off then jumped up, and chopped off a part of its head, but it kept coming for him. "Wow, these things are annoying!" Jack then sliced the rest of its head off, and that was it. They all took in a deep breath and relaxed.

"A few are not that hard to beat, but imagine thousands upon thousands..... But our magic should work. Looks like the head priest skinwalker is still sending out more and more after us." Lillith thought out loud.

Andrew looked sad; he wanted to fight. "Lillith, no matter what anyone says, I'm fighting with you.....okay?"

"Oh, I just thought you weren't ready, so I had to, thanks." He giggled and hugged Lillith tightly.

They all flew out of the forest and returned to DeVille.

****

When they got back to DeVille, they saw DeVille being attacked by a giant wendigo.

"Oh my Gods, let's save the city!" Lillith screamed at them and then ran towards the giant. They all ran towards it, taking out their weapons and starting to slice it, but Lillith's sword barely scratched it. But when Dom sliced its leg, it sliced through the leg, and it fell hard onto a house, but did not break it. It roared and turned its attention to the five.

"Why won't my sword work on this one?" Lillith took a few steps back, while the others sliced and destroyed the beast until its head was sliced off.

Everyone in the town ran out and thanked them all.

"It came and just started killing, that beast ate two poor elderly ladies, but you took care of it before it could kill us all!!" A woman came out and said happily.

Lillith approached her, “It’s our job, we are here to save the world.” Lillith smiled. “Is the pub still open?” She asked as her stomach gurgled.

“It is now, thanks to you guys. Come on, you are our honoured guests.” The woman spoke.

They all entered the pub and sat in the corner where they had for the past few days.

The waitress ran over, “What can I get you? It’s on the house!” She laughed.

“I’m kinda craving ribs....” Jack said quietly.

“We will have ribs, please, and six large cream sodas,” Lillith answered the waitress.

“Coming right up!”

After they ate, Lillith looked lost in thought.

“Why wouldn’t my sword work on the large wendigo?” She was obviously upset.

Jack cleared his throat. “I think it was because you are not using your weapon, your magic weapon, that wendigo was protected by magic, so your normal sword wouldn’t hurt it.”

“Fine, I will try to use my new weapon and just hold onto my sword and shield, then I will destroy all wendigos!" She took out her long sceptre and felt warm in her hand. "I will use this from now on!" Lillith announced.

Everyone at the table cheered and was more than happy to see Lillith happy.

They all relaxed, and Lillith put the map on the table. She pointed to the dark mountains, "We will rest here, at the bottom of the mountains, then fly over them, which are so high we will have to rest once we get over them. Then we will get to Crystal Henge after an hour. We won't have a town to go to that night, so we will have to sleep out in the wild. We will tough it out. We will bring food and drinks, so we will be fine. Does anyone have any objections?" Lillith asked.

"Nope, you are a great leader," Selene said with a smile.

"Kiss ass," Jack said.

"Oh, whatever, you brat!" Selene then slapped him on the arm.

After a while of just chatting and having a great time, Lillith stood up.

"Okay, let's head to bed," Lillith announced.

"Yes, Mom," Jack replied.

They all laughed and left the pub for the inn.

****

Lillith lay down next to Andrew and held him close. "You know, I just don't want to lose you, right?" Lillith spoke quietly.

Andrew smiled at her. "I just felt I had to take care of you, but I realised, you don't need protection.....it sucks being so useless." He closed his eyes.

"You support me and give me power by just having you by me, I don't think I could ever do this alone," Lillith whispered, then kissed him slowly. They held onto each other and fell asleep.

****

Lillith woke up, got ready, and then went to each room. All her friends were ready. They didn't have time to have breakfast, but they bought bread, cheese, beef jerky, and lots of water.

"Okay, guys, going south of here to the Dark Mountains, are you ready?" Lillith asked.

"We are!" They all said happily.

They got on their partners and flew south, towards the mountains they had to fly over. The mountains were really wide, so it would be faster to go over than around. It took longer than they thought to get to it, even more southern than where Lillith lived, but they reached it and landed at the base.

"Ugh, that was the longest ride we have ever been on." Complained Jack

"Yes, it was, and just think how tired our partners are," Lillith answered. She hugged her Praisia, who was still giant. She purred and knelt down so Lillith could get down. "Thank you."

They all got off their partners and sat in a circle as Jack made a fire with his kit.

They ate some of the food and drank some water, sharing with their partners. Then they relaxed, the partners slept soundly in a little pile they created of themselves, lying on each other.

"They are so cute!" Selene whispered.

"What to do for now?" Asked Jack.

"We prepare," Lillith said.

"What do you mean?" Jack was annoyed.

"Well, if the high asshole knew we were at that lagoon, he has probably put a tracking spell on us somehow, so who knows what is waiting for us at the crystal henge? He may know about that place. So I'm pretty useless right now. I will fly over the henge and jump off of Praisia into the centre of the henge, and you guys just get ready for a fight, I have a bad gut feeling about this place," Lillith commanded.

"Ohhh, more to slice up!" Jade laughed.

"Just be careful, okay, guys?"

They all nodded.

****

Praisia awoke first, she stretched and yawned, then started to wake up the others. They all stretched and flew to their partners.

Lillith smiled and pet Praisia. “You guys ready?” She asked the partners, and they all squealed.

One more thing.” Lillith turned to Andrew. “You are to wait here until we get back.”

“What?” Andrew exclaimed.

“I can’t afford to lose you, so you are staying here. If you don’t, I can’t protect you, so you are staying put. Please understand.” Tears formed in her eyes.

“Fine....” Andrew agreed but felt so powerless and useless. Then kissed Lillith.

“Okay, let’s get going!” Lillith said wearily.

They all flew off, leaving Andrew and Bob behind. The flight upward was hard. It kept getting colder and colder the higher they flew. Lillith looked at her compass nonstop. Ice started forming on their armour. But they kept going up, and the partners were not even affected by the cold. They finally reached the top, which was covered in snow and ice. Then they started down. They could see the forest below.

"We are almost there, guys, get ready!" They flew lower and lower than above the trees until there was an opening, and Lillith was right, there were four giant wendigos and four regular ones.

Lillith yelled. "I'm going to jump into the centre, I was right, the compass is shining lavender, that damn head skinwalker really is annoying me, you land over there." She pointed to a clearing a little away from the Wendigos. "I will jump in the middle, then you attack them, try and kill the large ones first, I will help as soon as I can!"

"Okay, boss!" Jack answered.

Lillith flew towards the centre of the circles of clear crystals and jumped down from Praisia and landed smoothly on the ground. Praisia shrank and flew to Lillith's shoulder. They were so dumb they didn't even notice her. She waited, then heard Dom yell, "GET THEM!!!"

Lillith sat on the flat crystal and hoped this wouldn't take long, just then she felt a surge go through her, the surge lifted her up her eyes opened and they shone lavender light, a wind the same color of her eyes that were glowing, surrounded her and then she felt a warmth she had never felt before, it was overwhelming her, but she didn't fight it, it poured through her like blood through her body. Her sceptre flew off her back as a large crystal formed on the empty spot at the top of her sceptre, in the shape of a blade, then a sharp edge on the bottom in lavender crystal. The sceptre was also dotted the whole thing with specks of crystal, making it even more beautiful. Then Praisias' head started to glow, and one large crystal formed on her forehead, and two smaller crystals formed on either side of her forehead. Then Lillith, as she was being bombarded with this magic, a crystal formed over the star on her forehead. So now the star that was thought to be a birthmark was now a crystal. Lillith was let down; she closed her eyes and felt so strong. She grabbed her sceptre that had been left her back, and screamed to her friends. "I'm coming!" And ran towards the giants that were attacking her team.

She jumped over the clear crystals and, with one slice, she chopped off one of the giant wendigos' heads, then ran to join her friends.

"Let's kill them fast, and not let one get away," Lillith told them. She jumped and sliced another head in half, then Jade sliced its head off. They killed one while she gained her powers, so one left. Dom cut through all the regular wendigoes. Jack jumped and sliced the head off the last giant.

They all looked at each other. They just killed all the wendigos like it was nothing! They were all so happy, laughing, shaking hands, and cheering. Then noticed Lillith wasn't joining in, and realised she was lying on the ground, passed out. Praisia was so worried, pacing on her arm.

"Lillith!" Selene screamed. They surrounded her, then relaxed. She was just sleeping. Then they remembered how tired they were when they got their powers. So she must have overworked herself. Dom picked her up and told the partners to grow. Praisia didn't listen.

"I think Andrew would rather I carry her than another man," said Selene.

"Yea, don't need to piss him off, here you go, I will hold onto her sceptre." Dom passed Lillith to Selene and then picked up the sceptre and hooked it onto his back. It was so heavy, he could barely hold it; it didn't like being held by him. "Wait, I'm gonna," Dom groaned. "I think this should stay with her. It's too heavy." He carefully put it onto Lillith's back.

"I think it's only heavy to anyone other than Lillith, because of the magic that created it and now fills it, just like the witch said, only we can use our weapons," Jade said.

"That sounds just about right." Said Jack.

Selene sat Lillith on her partner, who was now crouched down, then Selene jumped onto it and said. "Let's get Andrew!"

They all flew off, cheering that they finished this journey!

****

They landed with Andrew. Andrew looked worried. "Where's Lillith?" He yelled.

Dom walked towards him. She passed out. She was right, there were wendigoes there. She fought right after absorbing the mystic crystal and fell asleep.

"She's right here, Andrew, she is just fine." Selene carried Lillith to Andrew.

“My poor Lillith.....” He carried her to the fire and sat down, petting her head and noticing the crystal on her forehead. “Wow, you feel so powerful, but you always have.” He smiled and cradled her. “Well, the closest town is called Serie, we should go there after your partners refresh,” Andrew suggested.

“Sounds good to me.” Agreed Jade.

They all sat around the fire and explained all they did. Andrew was in awe. Everyone saw Lillith levitate, her eyes shone lavender, and they just didn’t go through that. But they knew after getting theirs, they were exhausted, so they could only imagine Lillith’s lack of energy, but she pushed herself way too far, as usual. But this journey will be over soon, just have to get back to the Elder witch, then who knows what. After an hour, they flew to Serie and put Lillith in a room, leaving a note on the table just in case she woke up. They tried to bring Praisia, but she wouldn’t leave Lillith’s side. They also told her where the pub was in the note, if she felt like coming. Lillith looked so peaceful, Andrew wanted to stay, but all of them made him come with them, so he did.

****

“There were four giant wendigos and four regular ones.’ Explained Jack.

“Yea, we destroyed them like it was nothing!” Jade said proudly.

“So, do you feel ready for this war?” Asked Andrew.

"Well, once we learn how to use our powers, we can take on thousands, probably more!" Dom bragged.

"I want to help......but how can I?" Andrew looked sadly at the table.

"We will protect you, so you are more than welcome to come." Said Selene.

"I dunno..... I feel like a burden."

"Only if you get in the way," Jack exclaimed.

"Shut up, Jack. Why are you such a downer?" Asked Jade

"I'm just thinking logically," Jack replied.

Bob rubbed on Andrew's cheek and purred.

"I'm going to go check on Lillith," he said quietly, left the table, and went to the Inn. He entered the room. He saw Lillith sitting up, eyes like lavender fire. He ran to her. She looked at him. She grabbed his shirt and pulled him close. She then kissed him. A surge of power entered him, and energy poured through him. A lavender crystal sword appeared, and then Bob, a lavender crystal, formed on his forehead. Just as Andrew felt he couldn't take any more, she let him go, and went back to sleep. Andrew felt a strange energy coursing through him. Was this magic? He was so tired, but he had to tell the others. Andrew ran to the pub and yelled at the group.

"Lillith gave me magic!" He showed them Bob and his sword.

"Wow." They all looked in surprise. "I'm sure the elder witch will know what happened."

"But what if you weakened her magic?" Jack said snidely.

"She forced me to guys, she grabbed me, I wouldn't have taken any of her magic! I never want to weaken her, I want her to be who she was meant to be....." Andrew looked down sadly.

"Jack, shut the hell up, obviously this was meant to be! I'm sure the Elder witch will know what to do!" Selena yelled.

"Oh man, I'm tired but hungry." Laughing, but still obviously upset. "Did you guys order yet?"

"Nope, but we are ready, what do you feel like?"

"Hm, I feel like beer-battered fish. And a large cream soda."

The waitress was flagged, and they all ordered. Bob flew onto the table and sat in front of Andrew protectively.

"It's okay, Bob, they are friends, even Jack." They all laughed. Bob snorted at Jack.

"Okay, okay, I'm sorry for being an ass, it's just the way I am, I'm sorry." Jack felt bad.

“It’s okay, Jack, I am nervous having powers when I’m not meant to.” Andrew smiled

Bob nodded and crawled back onto Andrew’s shoulder.

They wondered when Lillith would wake up. They all worried about her.

****

Lillith slept for two days straight. Andrew lay with her, loving her, hoping she’d be okay. Lillith’s eyes opened, and she stretched. Praisia squealed, and that woke up Andrew.

“You are awake!” Andrew laughed.

“What?” Lillith asked.

“You slept for two days straight, but we need to talk.”

“Two days? What do we need to talk about?” - Lillith

“While you were sleeping. I came and checked in on you, you grabbed me, then kissed me and gave me powers.” - Andrew

He showed her his crystal sword. “Even if Bob gained power, do you still feel strong?”

“I feel just as strong as I did when I got it. We need to check this out with the Elder witch, but I’m so glad you have powers. Now we can really take on these guys, both of us! Together!”

Lillith hugged Andrew and kissed his cheek. “You are my hero, don’t forget that.”

Andrew laughed and kissed her passionately.

Everyone was in the pub, and they all cheered when Lillith joined them.

“Let’s just eat, then go to my hometown, and then we can see the Elder witch. She will probably train us, no doubt, then it is onto war with evil, and save the world!” Lillith laughed, Prasia coed at Lillith.

Chapter 7 – Partners Forged in Fate

CHAPTER 7

PARTNERS FORGED IN FATE

Okay, let's head back to the Elder witch," announced Lillith "Praisia, can you fly up the mountain?" Lillith asked.

"Are you ready?" Asked Andrew.

"I am so ready, I feel like I'm going to burst!" Lillith laughed.

"I'm so pumped!!!" Said Dom.

"Then let's go!" Lillith said happily.

****

As they flew closer to the town, they saw smoke rising from 3 different places. Lillith's gut sank.

They landed and ran into town and saw three giant wendigos and a bunch of regular ones running around. They all took their weapons and attacked them. She saw one giant holding a girl, ready to eat her. Lillith jumped up and sliced through the giant's neck and then caught the child.

"I got you, don't worry." She placed the girl in her house.

They all attacked them, heads chopped off, then attacked the others. They stopped and made a circle, okay, Selene, Jack, and Andrew, you get the regulars, Dom, and I will take down the giants, we need to stop this attack!"

"Got it." They all listened to Lillith's commands and went all over. It was easy for them; they just kept slicing their heads off. It took them an hour to get rid of them all.

They looked around and saw many people who were hurt. She ran to one of her dad's friends. "Where's dad?"

The man looked up and frowned. "He led us, but we failed, and I don't know where he is..." The man said sadly.

"How bad are your wounds?" Lillith asked.

"I will live, thanks to you guys. Your dad was separated from us, but the attack only just began when you came, so I don't think anyone even died." He smiled as blood dripped down his face.

"I suggest you see the Elder witch and bring anyone else that needs help," Lillith told him.

"Yes, I will gather up all the men, and we will go." The man agreed.

Lillith ran all over. "Dad? DAD!!!! Where are you?" All the fires slowly died down, and she heard coughing from one of the broken-down houses. It sounded like her father. Lillith ran into the falling house and found her father carrying two children to get them out of the crumbling house.

"Dad!" Lillith laughed and helped him with the children, and as soon as they stepped out of the house, it collapsed. Her father was all bloody and hurt pretty badly. "Dad, you need to see the Elder witch to help you heal, with no infections!"

Lillith's father let the kids go, and they ran to a neighbour's house.

"Okay, did you guys take out all those monsters?" Her father asked.

"Yes, it was easy, but by far the most we have killed at once. The secret is to chop off their heads."

"Oh, well, that's hard to reach the gigantic ones."

"I guess we can jump higher now since we gained our powers," Lillith admitted. She put her arm around her father to help him walk.

"What's that on your forehead?" Her father asked.

"It's a crystal," she explained.

"It suits you, hun." He tried to laugh, but his chest hurt too much.

"Calm down, we are almost to the Elder witches' hut."

"Good, because I feel dizzy," said Lillith's father.

Lillith banged on the Elder witch's hut.

The elder witch opened the door, and a huge smile grew across her face. "You are back!"

"Yea, just in time to save most of the town. But please take care of my father, he is badly wounded." Tears welled up in Lillith's eyes.

"Oh, yes, dear, bring him in." She opened the door and led the father to a table; he lay down on it."

She ripped off his shirt; there were five deep gashes, he had lost a lot of blood, and he passed out.

"Oh my Gods!" gasped Lillith.

"Don't worry, my dear, thanks to you, no one died, you and your team. Now let me stitch him up; no arteries were cut, he just passed out because of the blood loss, but he will be fine." The Elder witch said, trying to calm Lillith down.

"Okay, work your magic," Lillith said calmly.

The Elder witch sewed the torn muscles, applied healing salves, and then sewed up the largest wound. Then she put a healing salve on top and started to work on the others. In no time, he was all patched up.

“There, your father is now ready to heal. He should stay in bed for at least a week.”

“Okay, but how do we wake him?” Lillith worried as people started knocking on her door.

“Oh, that is quite easy, my dear. The Elder witch took a large bucket of water and poured it on his face, and he jumped up.

“What the?” The father yelled.

“Now take him home and to bed,” the Elder witch told her kindly.

“Okay, we will come back when things calm down, okay?” Asked Lillith.

“Sounds good. How about later this evening, like after dinner?” The Elder witch wondered.

“Sounds good, we will see you, oh, and something strange happened, but we will talk about that later. See you soon.” Lillith left the shack, passing injured people, but none as hurt as her father. She walked her father to their house. Mother hugged them both and helped Lillith get her father into their bed. “Father, you are bedridden for a week, understand?” Lillith ordered.

“A week?” He pouted.

“Yes, and mom will make sure you do! Right, mom?” Lillith asked.

“Yes, I will.” She told both of them.

“Mom, I need to find my friends, I’m sorry.......I have to save the world and am running out of time.” Lillith said urgently.

“Yes, go, my dear, my heart is always with you.” Her father said.

They hugged, and Lillith ran out the door.

****

She looked in the pub, and they were all there except Andrew.

“Where’s Andrew?” Lillith asked urgently.

“He went to see his mom and dad.” Answered Jade.

“Okay, I will go look for him. You guys have fun, you deserve it.” She winked at them all and ran out. She ran all the way to Andrew’s house, then knocked on the door.

The door opened, and she saw Andrew’s mother. “Oh, my future daughter! Please come in.”

Lillith entered the house and looked for Andrew.

“Is everyone okay?” Asked Lillith.

Oh, Andrew's father is fine. He had broken his arm trying to fix the window, so he couldn't fight. Thank the Gods for that."

"Is Andrew here?" Lillith asked.

"Yes, he is just going to the bathroom." She giggled.

"Oh, okay." As soon as he is done, we have to leave. We have to talk about important things." Lillith told her future mother-in-law.

"Oh, of course, we are just so proud of you guys, killing all those THINGS, like it was nothing!"

"We will be fighting hundreds and hundreds, so twenty or more is pretty easy," Lillith explained.

"Oh my, but you two are very well trained, I know you can do it!"

"Thank you." Lillith smiled, then Andrew left the bathroom.

"Do you feel better?" Lillith laughed.

"Much!" Andrew laughed.

"We need to go to the pub to talk, let's go!" Lillith grabbed Andrew by the collar and dragged him out.

****

Lillith and Andrew entered the pub and smiled at their friends.

"Okay, as we flew here, I didn't see any other town around here attacked. That bastard found out where my birth town is, he is trying to destroy me, but he will be the one destroyed!" Lillith growled. But we are to see the Elder witch after dinner. What time is it now? Does anyone know?" Lillith asked.

"It's a little past four," answered Jack.

"We saved this town, guys! Thank you, and no one died!" Lillith exclaimed.

"We are really a great team, and I think the new addition is really helping us out." Lillith pointed to Andrew.

Andrew smiled. "I'm honoured I can help."

They all laughed and cheered. They ordered a roast beef, so they can all share, with delicious gravy, and of course, cream soda.

They ate and had a great time, feeding and sharing drinks with their partners. They wondered how much more they would have to do to be ready. Now that's up to the Elder witch. She has to train them, so they can fight hundreds and come out the victors. Also, their partners need to learn their powers, which will help them tenfold.

"I have a question for you, Lillith." Asked Andrew.

"Yes?" Lillith looked at Andrew.

"Were you aware of giving me powers?" Andrew asked.

"Honestly, no, I was sleeping. I guess the magic felt you were worth it." She smiled.

"Okay, that feels better. The magic chose me." Andrew smiled.

****

"It's seven, we should head over to The Elder witch's hut," said Jack.

"I agree," Lillith said, and they all left, excitedly walking to the Elder witch's hut.

They knocked on the hut's door.

The elder witch opened up, "So you gathered all the elements?" The Elder witch asked.

"Yes, but something surprising happened," said Lillith.

Lillith stepped out in front. After I gained my powers, I passed out for two days. During this time, Andrew was given magic by me, but I was out cold. I did not give it to him. I'm glad it did, but will everything be okay? Lillith asked, a little worried.

The witches' eyes widened, and they laughed. "The magic chose him, too. You six now must defeat the evil together instead of just the five of you!"

"Oh?" Lillith was glad.

"You and Andrew must really be close, you two are kindred souls, you both have the same destiny, the magic chose him." The Elder witch said proudly.

"Does it mean I lost some power?" Lillith asked.

"Oh no, no no, you've added to it, Andrew, can you come here?" Asked the Elder witch.

Andrew stepped next to Lillith.

"Oh yes, you both are very powerful, all of you are strong!" The Elder witch said happily. "All of you are radiating power; magic is in you now and forever. Okay. The training starts tomorrow morning, so eat a big breakfast, and be prepared to work your butts off!" The Elder witch laughed and went into her hut, giggling happily, then closed the door.

They all looked at each other and shrugged.

"I thought we would be starting tonight..." Pouted Jack.

"I can't wait to start tomorrow!" Jade, Dom, and Selene said in unison.

"Okay, let's get rooms at the Inn; you may have to share rooms, so be prepared," Lillith announced.

****

They went to the inn and got three rooms: one for Selene and Jade, one for Dom and Jack, and one for Lillith and Andrew.

“Good night, guys,” Lillith said, then entered their room.

Andrew closed the door, then walked up behind Lillith and hugged her. “To think, we are going to be fighting a big battle. It makes me feel good inside, like we were meant to be.” He whispered in her ear.

Lillith smiled and turned around, and caressed his cheek. I’m not surprised, I’m relieved, you always give me hope and power, you are my soulmate, my love.” She whispered back to him.

Andrew held her close. “I have never felt so complete in my entire life. I am so lucky to have you.” He rested his head on Lillith’s head.

“One day we will get married and we can start a family, you are my dream come true.” Lillith thought.

“I love you, Lillith, with all of my heart,” Andrew said softly.

“And I love you, my darling. Let us go to bed; we have a big day ahead of us.” Lillith advised.

“Can we hold each other while we fall asleep?” Andrew asked.

“I would love that.” Lillith giggled and jumped into their bed.

They cuddled and fell asleep as Praisia and Bob slept together on a comfy chair.

They didn't realise how exhausted they were and passed out.

****

Lillith woke up first. She felt so warm in Andrew's arms. She could hear his heart beating, and that made her feel so happy. Praisia noticed Lillith was awake, so she flew onto Andrew's head and messed up his hair to wake him up.

"Praisia, that is not nice." Lillith laughed as Andrew started to wake up. Praisia then hopped onto LiIlith and licked her and squeaked.

Bob woke up and joined the three and squealed at Andrew. Andrew lifted a hand to show he was up.

"Good morning, my beloved Fiance." Lillith giggled.

"Good morning, my dream has come true." They hugged, then kissed. Andrew then got up.

They got ready slowly, both taking a shower and the usual. Then headed to the pub. They would wait an hour before waking the others. As they walked into the pub, they were surprised to see everyone there.

They all chatted, then ordered, and were so excited for today. Today was going to be a hard one, but who knows, it might be fun too.

"Okay, time teller, what time is it?" Lillith asked Jack.

“It’s 9:15 am,” Jack replied.

“Are you guys ready to start training?” asked Seline.

“You bet!” Laughed Jade

“Okay, let’s go!” Lillith cheered.

****

They knocked on the Elders’ door and waited.

She was at the door in a flash. “Okay, let’s start the training. Let’s go into the field for more room. I want you all to drop your weapons right here. not in a pile but next to one another.” Ordered the Elder witch.

They did as the Elder witch told.

“Now walk forty steps away from your weapons.”

They did, but were confused.

“Now I want you to bring your weapons to you with magic and nothing more.”

They all raised their hands and tried, but failed. They tried for ten minutes and still nothing.

“Can you give us some clues as to how to do it?” Asked Lillith.

“I was waiting for that. First, feel the energy the magic has given you, feel the warmth. Then, remember the feeling of the power of your weapons, and connect to it. Then, call it. It should come; it’s your weapon. It’s a part of you.” The elder explained

As the six pictured their energy, they began to glow. They all raised their hands and felt a pull. They then pictured the weapon being pulled towards them, and all their weapons came into their hands. And they cheered as they passed, and how simple it was to do.

“Now that wasn’t so hard, now was it?” the Elder witch giggled.

They took a fifteen-minute break while the Elder witch prepared for the next thing to learn.

They were lined against six wide trees.

Now you need to learn how to attack using magic, look deep inside yourselves, feel the warmth, put it into your weapon, and attack. Now practice!” Ordered the Elder witch.

They all tried and failed for the first few times, but then they started getting it. It took them a while, but they chopped down the six trees. A tree each. They all cheered.

“No, not good enough yet, you should be able to chop a tree down in one strike!” She smiled, “This may take a few days to learn. See how you swing your weapons, and a magic slice comes out? It should be bigger and stronger; it should take out multiple enemies at once." The Elder witch said sternly.

They looked at each other, then nodded and continued to practise. They practised for hours, and the Elder witch came out with sandwiches and milk.

"Take a break, it is all about picturing it. If you wish it to slice through the tree, it will," the elder witch advised.

They all took a sandwich and ate, then had some milk.

They then tried picturing it and willing it to cut through a tree. Eventually, they all were getting stronger, and soon after that, they cut through a tree like it was nothing.

"Very good! Now, cut down five at once, they all moved around, and soon enough, they were cutting down six at a time.

"Well, isn't that even better. Your skills are really getting good fast. Now, to make a shield to protect yourself. Picture your magic surrounding you in a crystal ball; nothing can break that. If you master this, you are pretty much done with the fighting part." The Elder witch laughed.

They all pictured it, and they all were surrounded.

"Ok, I'm going to throw magic at all of you, to see if your shield is strong enough." She lit a match, then made the fire hit everyone, like a wild flame surrounding them. She did it for a minute. She stopped and checked on them.

Lillith was wide-eyed. “I didn’t even feel the heat!” Lillith said, shocked.

“That was quick! You guys are great! Okay, now I’m going to teach you to use your elements. Jade, you first.”

Jade walked over to the Elder witch and waited.

“Okay, picture flames, then make it appear in your hand,” ordered the Elder witch.

Jade put out a hand. She was getting used to picturing magic now, so she pictured it and brought it to her hand.

“Good job, now will it to strike me.” The Elder witch ordered.

Jade pictured a flame shooting out of her hand, attacking the Elder Witch. It bounced off the Elder Witch like nothing.

“Good, now think bigger.” The Elder witch smiled.

Jade pictured huge flames, then shot them at the Elder witch. Bounced off.

“Why do they not hit?” Jade asked, annoyed.

“Well, if it hit me, I would be dead, so they are successful, so be happy, you can make the fire do anything." The elder witch told Jade.

Elder witch pointed to Dom."Now, Dom, let's see what you got."

Dom put out his hand and pictured a swirling ball of water. It appeared above his hand. He was then ordered to use it against the Elder witch. He pictured it turning into a spike of ice and shot it at her. It broke when it hit the barrier of the Elder witch.

"Very good, Dom! Now, Jack, let's see you work your element."

Jack pictured a small tornado in his hand, and it appeared, then shot it at the witch, but made it much bigger. The elder witch clapped as the wind bounced off her.

"Good thinking, Jack," She then pointed at Selene, "Let's see your work."

Selene opened her hand and made a spike-like object appear, then elongated it and made it fly at the Elder witch. It broke into splinters as it hit the Elder witch's barrier.

"Wow, creative Selene. Now you, Andrew."

Andrew made a lavender sphere, then transformed it into a blade and shot it at the Elder witch. It stopped in mid-air."

"Oh my, you penetrated my shield, luckily I stopped it, or I would not have survived that blow! Great job. Now, Lillith, you are to aim at the rock, because I know I won't be able to stop the thing you make."

“Okay...” Lillith formed a star, then turned it into a huge spike and shot it at the boulder; the boulder exploded. Lillith looked shocked.

“Well done, my dear. I’m not shocked. Why are you?” Asked the Elder witch.

“I just don’t understand why mine is so strong?”

“Your element is the strongest, that is why.” She smiled at Lillith. “You are the leader, you are the chosen one.”

O.......kay,” Lillith looked at Andrew, who smiled at her.

“Okay, my lovelies, you passed with flying colors. I figured it would take at least two days, but you guys are great! I made special armour for you guys. Put that on and you will be ready to start fighting this war! Oh yes, and your dragons have power too; they shoot it through their mouths, but I bet your little partners knew that already. “ Elder witch explained.

They all flew around happily. They knew how to, and they were smart.

“Praisia, chop down that tree!” Lillith ordered.

Praisia snorted and shot out a big burst of lilac flames, and it turned the tree into soot.

“Oh wow, good job hun!” Lillith praised Praisia.

They all got their special armour, which was made out of tough leather, thanked the elder witch, and went to their rooms, where they were to meet up at the pub.

****

“Are you guys ready?” Asked Lillith.

“I am,” said Selene.

“Now that we have power, I’m ready to get rid of the damn evil taking over our world,” said Dom.

“I’m ready to get rid of evil, too,” said Andrew.

“I think we all are ready, after all the crap we have been through,” said Jack.

“I wonder how many there will be. Do you know where they are?” Asked Jade.

“I think if we fly, we will see where they are; there are supposed to be hundreds upon hundreds.”

“I think the hard part will be to find the head skinwalker. I think if he sees us fighting and killing his wendigo, he will escape, so we might have to come back here to ask the Elder witch for help.” Lillith pondered out loud.

“That will probably happen, unless he thinks he can take us, I mean, he has got the dark crystal.

“We have to get the crystals all together to make the special crystal.” Said Andrew

“Oh crap, we need to ask the Elder witch how to do it.....I need a refresher on that.” Lillith pouted.

“That is fine, we will all go tomorrow, she will tell us,” Jade answered.

“I should have asked today; now we have to wait to leave,” Lillith complained.

“We learned so much today, so please don’t feel bad.” Selene comforted Lillith.

Andrew put his arm around Lillith. “We are not behind, we are ahead, so don’t worry.”

Lillith took a big gulp of her cream soda and smiled. “Okay.”

“We could also ask where we will find the army of wendigo and how to track the skinwalkers.” Jack pointed out.

“That’s true.” Lillith smiled, “She might know how, then we can destroy the wendigo, then take care of the bastard skinwalkers. Especially the head one, I bet he would hide from us.”

“Exactly, so don’t feel bad, okay?” Dom pointed out.

“Okay, so we will ask after breakfast, okay, guys?”

“Sounds good,” they all said.

“Hey, we might have to save a few towns before we even get to the army.” Jack pointed out.

“Very true.” Lillith agreed.

Jade yawned, and then they all started yawning.

“I think we are all wiped out. Let’s go to bed, have an early start tomorrow,” Lillith said with a weak smile.

Praisia flew onto the table and was joined by the other partners. They all watched as Praisia walked around them, nudging all the others. Bob squealed and rubbed against his mother, then they all flew back onto their partners.

“Looks like Praisias is the leader. She’s preparing the others for our journey. to the war we have to fight.” Noted Jade.

“Praisia, you guys don’t need to help until we fight the army, so save your energy.” Lillith directed Praisia

Praisia nodded and nodded towards all the other partners.

Merlo nodded and called out in a war cry, but quietly.

Magna and Umi bowed their heads, and the rest all bowed to Praisia as well.

Bob flew to his mother and gave her a good rub, then flew back to Andrew.

Praisia purred, then fell from Lillith's shoulder. All the partners fell, and they all caught them.

"Crap, they are getting sick again! The evil must be growing!" Lillith pouted. She went to her side pack, searched, and fed each a nut that would help them. They all ate a nut, but they were really hurting. The nut helped, but Lillith could tell they were still hurting. She focused on them and could see a dark purple aura around them.

"That bastard Head skinwalker is cursing them!" Lillith pictured a protective aura around herself and Praisia. She curled up and coed; the dark aura vanished around Praisia. "Guys, picture a protective aura around you and your partners! Quick!"

They all did, and then the dark purple aura vanished.

All of them hugged their partners, who were happily rubbing against them.

"Thank you, Lillith. How did you know?" Asked Jack, who was now babying his little Merlo.

I saw the dark purple aura around them. Couldn't you guys see it?"

"No," They all answered.

"That bastard used the dark crystal to kill our partners!" Lillith kissed Praisia and held her close.

"We have to be careful; if he can curse our partners, could he use that energy against us?" Asked Selene.

We just need always to keep picturing an aura around us and our partners, even if we are separated. Just keep it up in the back of your minds. Seriously, I see the auras you guys put up, all your colors. You guys can't see them?" Lillith asked.

"No, thank the Gods you can!" Said Dom.

"I have no idea why I can? I don't understand...." Lillith wondered.

"I guess we have another question for the Elder witch tomorrow," said Andrew.

"Lillith wondered why she was different, why she was stronger, why she had this star on her forehead that was now a crystal. Lillith hoped the Elder Witch could answer these questions.

"Whatever the answer is, Lillith, I will always love you." Andrew smiled but looked worried.

“Why do you look worried, Andrew?” Lillith asked.

“I’m just afraid I will lose your love if you find out something that will change your mind after you find out,” Andrew answered.

“Andrew, nothing and I mean nothing can ever change my feelings for you, I love you now and forever.” Lillith smiled and took one of his hands and kissed it.

“Okay, good, because I just can’t go on without you.” He leaned in and kissed Lillith softly.

“You two are so cute.” Selene smiled.

Jack sighed and looked annoyed. “Too much, you are showing too much love!” Merlo rubbed against Jack’s hand.

“Sorry, guys, we will stop.” Lillith and Andrew looked embarrassed.

“Don’t be embarrassed!” Ordered Jade. “Jack’s just jealous.” She stuck her tongue out at Jack.

Jack rolled his eyes.

They all giggled at Jack.

“Okay, let’s finish our drinks, then go to bed, big day tomorrow,” Lillith said, then chugged her soda.

They all left and returned to their rooms. Praisia flew to Bob and then flew to the comfy chair next to the bed. I guess they want us to cuddle." Andrew laughed.

Lillith pulled Andrew to the bed and cuddled him. "We will be together forever." She kissed Andrew on the neck, then relaxed.

"I wish this good feeling would last forever," Andrew admitted.

"It will," she whispered, then closed her eyes.

"It better." Andrew laughed and relaxed. They both fell asleep in seconds.

****

Andrew woke up for the first time and felt Lillith's warmth. This was heaven. He refused to move because he didn't want it to end. But Bob noticed Andrew was awake and flew onto Lillith's shoulder and licked her face.

"No, Bob, let her sleep!" Ordered Andrew, but then Praisia followed and rubbed against Lillith's neck.

"You guys are mean." Andrew laughed. "Poor Lillith."

Lillith stirred and slowly woke up.

"I'm sorry, hun, they are just excited, I'm guessing," Andrew said with a smirk.

“It’s ok, I was just dreaming of us getting married.” She laughed and kissed her little Praisia, then pet Bob, who was now jumping on Andrew’s shoulder.

“Lillith.....let’s get married after we get all this stuff done...I mean, it would be great to have that to look forward to.” Andrew smiled.

Lillith giggled. “It’s about time we set a date, don’t you think?” She joked.

“I don’t know why we waited, but it will be worth it.” He laughed.

“Yes, it will.” Lillith agreed, then snuggled against Andrew.

“It’s still dark out, we have lots of time to cuddle my sweet.” Andrew rolled up against Lilith and petted her head, then noticed something. “What the?”

“What?” Lillith asked.

“Your hair is turning lavender, like your eyes.” He looked at it, and it was almost down to the tips.

“WHAT?” Lillith said, annoyed with her hair. Lillith jumped up and ran to the bathroom. Her hair was indeed turning lavender. “I liked being a redhead.....” She pouted.

“I think it looks stunning. Lavender was always your color.” There were still four inches of her old hair color left.

“I think I’m changing.... but into what?” She wondered.

“I think you were always meant to have lavender hair; it makes you look more mystical. Maybe it has to do with who you were meant to be?” Andrew wondered out loud.

“I wonder what I am.....” Pouted Lillith

“Whomever you are, you won’t be anything but my Lillith.” He pulled her back to bed and then kissed her head.

“Well, at least I match more.” She giggled, running her hand through her silky hair.

“You do, and I wouldn’t change a damn thing about you!” Andrew said and pulled Lilith into a soft kiss.

CHAPTER 8 – THE PRICE OF A DYING WORLD

CHAPTER 8

THE PRICE OF A DYING WORLD

Lillith and Andrew decided to take showers and get ready. They took their time because the pub didn't open until 8. They didn't have anything to tell the time, so Lillith and Andrew took their time. They decided to take a walk, so they walked around the town. The sun had already risen, so they went to check and see if the pub was open yet. The lights were on. They pushed on the door, and it opened. Lillith stuck her head in. They heard the waitress say, "It's not open yet." Lillith was about to say, "Sorry," when the waitress noticed who it was who peeked in.

"Oh, Lillith! You can come in, you saved my dad, come in, come in." The waitress smiled at her.

Lillith smiled and tugged Andrew to enter, "Sorry, we woke up early."

"That is fine, you are the heroes, there will always be an open door with you and your friends."

"Do you know what time it is?" Asked Andrew.

"Oh, let me see, it's almost 8, so you are not that early, don't worry, if itwas early, early we would have had the doors locked, so go take a seat, the cook is here if you want breakfast, he will cook it."

“Oh, we will wait for our friends, but we would like two glasses of cream soda,” Andrew replied.

“Make that 3, one for our partners,” Lillith giggled. They deserved a treat.

“Okay, go take your seats, and I will bring them to you.” Smiled the young waitress.

Lillith took the table in the corner that would fit them all. She sat down, and Andrew sat next to her.

“Let’s let Praisia and Bob play on the table,” Lillith suggested.

“I think Bob would love that.” Andrew smiled, then picked up Bob and placed him on the table. Bob stretched and then yawned.

Lillith placed Praisia next to Bob, who started to bug Praisia by hitting her with his tail. Bob then tackled her, and they rolled around. They were about the size of Lillith’s hand, so they didn’t even move the table. Lillith and Andrew giggled.

“They are such sweet creatures.” Lillith Laughed.

The waitress came to them and put the drinks on the table.

“Cute lizards.” The waitress laughed, “I’ve never seen a gold and silver lizard before.”

"Yes, they are rare." Andrew laughed.

Lillith picked up Bob by the scruff of his neck and scratched his belly lightly. Bob squealed. Praisia lightly bit Bob's tail and pulled him around. Lillith placed him back down, then he rubbed against his mom, Praisia. Praisia started to lick his face. Bob didn't like this, but let her. If he refused, she'd kick his butt, so he had no choice.

Lillith picked up Praisia and showed her and Bob that the third cup of cream soda was theirs, and she drank it and burped. Lillith laughed as Praisia burped. Bob took a big gulp and also burped. They finished off the soda.

Lillith called the waitress and asked for two bowls of milk. The waitress brought some, and they drank them down too.

There was nothing left to do but wait, which was okay, because Lillith and Andrew talked and held hands.

"You know, I'm not human, right?" Lillith asked.

"Close enough." Andrew laughed.

Lillith kissed him on the cheek and leaned on Andrew.

****

Selene and Jade walked into the pub and sat at the table. Their partners joined the two at the table.

"So the guys are sleeping in?" Selene smirked.

Jade laughed and ordered them both a cream soda.

Dom and Jack came in ten minutes later and joined them. Lillith ordered them both a cream soda.

They also all ordered breakfast, and they ate it all up, sharing with their partners.

"Okay, we need to go to the Elder witches and get our answers," Lillith said.

"Aren't you nervous, Lillith?" Asked Dom.

"Very," Lillith answered, then looked at Andrew. Andrew squeezed her hand.

They all got up and left for the Elder witch's hut.

****

They walked to the Elder witch's house after breakfast. Lillith went to knock. The door opened, and the elder witch stepped out with a big smile on her face.

"I knew you were coming. How can I help?" The elder witch asked.

Dom stepped forward, "We have a few questions." Before Lillith could say.

“Then ask away.” The Elder witch smiled.

Lillith stepped forward, “First, how do we know where evil is?”

“I have a gem for that! The wendigoes show as green, and the skinwalkers show as red. And when you are approaching either, it will let out a vibration that you will feel in your hand.” She reached into her pocket and out came a blue crystal, large and clear enough to look through.

“Oh, also, there are five different camps of wendigos, one for each skinwalker, then you will find the head skinwalker once you defeat the others. But I suggest you come back here to refresh yourselves, then I will help you locate the head skinwalker.” The Elder witch told.

“Okay,” said Lillith, and they all nodded.

“How do we combine our elements?” Lillith asked.

You and your partner stand in the middle, and the others surround you, and you all picture it combining into one big crystal. Picture your element entering the crystal of Lilliths.”

“That seems easy. “Jack said.

“Actually, it takes all your energy, so you will probably all pass out. Your partners will protect you, and you should do it when you defeat all the wendigos and the five underlings of skinwalkers. Skinwalkers tend to be tricky things.” The Elder witch explained.

Jack shut his mouth.

Lillith then gulped. "What am I? Not human?"

"You are of the elder race, bonded by magic, a Unicorn, but it's just a name because of the marks you are all born with. Like a separate race of elves, but you are much more special and more powerful. You are meant to save the world."

But why me, why not another?" Lillith questioned.

"Most unicorn people are born with a five-star birthmark. It was said that a sixth star would come and free the world of evil. And you were dropped off in my very town, no travelling for me to wait for you." She booped Lillith's nose.

"So I'm a horse?" Lillith asked.

No, darling, you are an old race that just used the name because of the markings of a star on the forehead. You have evolved over many, many centuries. But you are not of the mythical beasts."

"Oh." Lillith looked at Andrew. Andrew put his arm around Lillith.

"This was a prophecy to happen, and it did, and you are here now! We are so grateful because you and your friends are meant to save the world and bring magic back to our world!

Lillith closed her eyes. Tears started appearing, and she wiped them away. “I need to see my parents!” Lillith then ran all the way to her house, leaving her friends behind. She thought about how strange she must be to all her family. Why must she be different? Lillith thought. She ran up to her door and opened it, with tears still in her eyes. She ran in, and her dad was still in bed, with her mother by his side. The two woke up.

“You wanna know what I am?” Lillith asked.

“You are not human?” Her mother asked.

“No, I’m called a unicorn, but not because we came from unicorns, it’s because of the stars on our foreheads, on the unicorn people, it’s normally five star points, but mine has six, so I was brought here to get rid of the evil, I was destined to save our world. There was a prophecy, and they thought you guys would raise me properly, and obviously you have.” She laughed. “I think you two were part of the prophecy, too.”

They both hugged Lillith.

“I will always be your daughter, no matter what happens, understand?” Lillith told them.

“Of course, my angel!” The two parents said to Lillith, her mom kissed her daughter’s forehead.

“How will you defeat the evil?” Asked Lillith’s father.

“With weapons like this, Dad.” She pulled her sceptre from her back.

It flowed with powerful energy, and the purple crystal radiated now that she was holding it.

“Wow!” Her mom whispered.

“Cool, hmm?” Lillith laughed a little.

“Your hair has changed too, beautiful, with some of the old color at the tips.” Lillith’s mom pointed out. “You look beautiful, angel.”

“Really, you think?” Lillith looked shocked.

“Yes, you are the most beautiful girl I have ever seen, besides your mother.” Lillith’s dad commented.

“I just want to thank you for everything. I wouldn’t have learned anything if I were in a regular home. I wouldn’t have met Andrew, and you two have always been supportive. I want to say thank you. I will never be able to pay either of you back for everything you have done for me. I love you both so much, I can’t express how much I love you both. Now I have to go save the world.”

Both parents were crying happy tears. “We want you to know how grateful we are to have you in our lives. Go save the world, sweetheart, you are in our hearts, and never give up!” They all hugged one more time, then Lillith ran out the door.

“Oh shit, I forgot to collect the crystal to help us find the evil bastards. She walked down the street, wondering where everyone was. She decided to just go to the pub and wait for them. She entered the pub, and Andrew was sitting waiting.

"Andrew...." She ran and hugged him tight. She smelled him; he always smelled of lavender to her.

"Hey sweetie, are you okay?" Andrew asked, concerned, and he started rubbing her back to soothe her.

"I need you always, Andrew. We are meant for each other, and we are going to live life together forever. Right?"

"I have been by your side since I met you, at seven and you were five, I will never leave your side, I got your back." He smiled and they kissed in a long, passionate kiss.

Lillith took a deep breath, holding onto Andrew's hand. "Where is everyone?"

"Oh, they will be here soon, they are talking with the Elder witch, she's giving them ointment and so on, they are bringing a lot of stuff just in case," Andrew explained.

"Why aren't you with them?" Lillith was wondering why they'd separate.

"Well, I followed you and watched you get to your house, then I thought...." thought, where would Lillith go after saying goodbye to her beloved family. I believed it would be here, so I've been waiting for you, just in case you needed me. You are the most important person in my life, so I always have to protect you." He said with a single tear running down his cheek.

"Oh my poor Andrew, why are you crying?" Lillith asked.

"Because I'm worried I might lose you, or you might lose me in this fight...." Andrew admitted.

"Oh...." Lillith bit her lip. She never thought that could happen.

"Yea, it will break my heart if you die, and it will break your heart if I die. Is this some cruel joke the Gods made for us?" wondered Andrew out loud.

"If either of us dies, I want you to move on. You are an amazing man, and you would be a great dad. But always know I will always be by your side, and no, I won't get mad if you find somebody." Lillith smiled, but her eyes said she was sad.

"Hun, I will never marry or fall in love again, because I've had the best and I don't want the rest. I want to always be loyal to you, in life and in death. We will be soulmates, and one day I will meet with you on the other side," Andrew replied.

"Then I will do the same, because I don't ever want to love another; I want to be loyal to you."

"Okay, let's make a pact: if one of us dies, we must fight the hardest to survive, so at least one of us can live a long life," said Andrew.

"I agree to it, I can't even picture cuddling with another man but you, you will be by my side while we are fighting. I don't want to lose track of you, because I am going to protect you, so are these two." She pointed to Praisia and Bob. They both purred and nodded.

"You two take care of Lillith, too, ok? You are our guardians?" Andrew begged the two dragons.

They both nodded.

Lillith laughed, and Andrew started laughing because Lillith was laughing. They felt so much better.

"Let's dance, we don't need music to dance," Lillith said, blushing.

"Slow dance?" Andrew asked.

"Yes, of course." They both got up. Lillith placed her arms around Andrew's neck and rested her head on his chest. Andrew wrapped his arms around Lilith's waist, and they just danced, saying sweet nothings to each other. They danced until their friends entered the bar. Lillith was feeling so much love for Andrew that she felt like she could explode. Knowing Andrew was feeling the same way, they were the perfect couple; they were soulmates.

"So what's going on here?" Asked Dom.

Lillith smiled, "Just making a pact, but don't ask what it is, because we will never tell." She winked at Dom, Lillith took Andrew's hand, and led him to the table. They sat extra close to each other. Their friends looked at them, "What the hell has happened to the two of you?" Asked Jack.

"For the first time since all this stuff started, we are at peace of mind, and it's just so beautiful."

"Geezez, I will have what they are having!!" Announced Jade.

"Yea, me too!" Said Dom.

"Sign me up!" Yelled Selene.

"Everyone laughed.

"Okay, everyone, tell your ages," Jack asked to change the subject.

"I am 18, Andrew is 20, Dom is 24, Selene is 20, Jack, you are 26, and Jade is 21," answered Lillith.

"Let's just relax for the night and head out tomorrow after lunch?" suggested Lillith.

"Sounds good to me," said Dom.

Remember, you four, you are fighting evil with your partner, you each will have each other's back,

All the dragons chirped, including Praisia and Bob. Everyone laughed.

“Come on, guys, let’s get some food! And, I think everyone should be allowed to have one beer to celebrate our fighting spirit,” said Andrew.

“AGREED!” Everyone said, including Lillith.

“Ok, medium rare steaks for us all!” They all ate their stakes and drank the beer.

“So, Dom and Selene, you will attack from the right, Jack and Jade will attack from the left, and Andrew and I will attack from the middle, sound good?” Lillith asked.

“Yes, sir!” they all agreed.

They decided to let their dragons play with each other on the table. They all were having a great time, the dragons and themselves.

They all agreed.

****

Lillith lay in bed and just couldn’t sleep.

“Hun, are you okay?” Andrew asked.

“I’m just worried about our fight and how tomorrow we are going head-to-head with them," Lillith explained. She smiled at Andrew. "To be honest, I'm worried about tomorrow, I mean, you know I love to fight, but what if we get tired halfway through? The wendigos won't wait for us to get our breath back?" Worried Lillith.

"Well, if we run out of steam, we can fly away and come back in the morning, then take another break until every last one is dead."

"That's a good idea, but I think we will get through them all, but if not, it doesn't mean we failed; we just need to destroy them."

"Don't worry, sweetie, everything will work out."

Thank you for listening. I feel much better, and I think I can sleep now." She wrapped her arm around Andrew's chest and cuddled. Soon they were both sleeping.

****

The next day, Lillith woke up, stretched, and got ready to fight. She woke up Andrew slowly so he would feel good waking up.

Andrew woke up nice and calm; he was ready to fight his hardest. They both took a shower, put on the armour that was made especially for them by the Elder witch. It fit perfectly and was comfy and light. They packed up and left for the pub. When they got there, everyone was there and they were glowing with happiness, wearing their armour too.

"So it looks like you guys had a great night's sleep," Andrew announced. And looked happy for them.

"Let's order a big breakfast!" Lillith cheered.

They all got three eggs, seven sausages, four pieces of bacon, and two toasts. They shared their food with their partners.

"Waitress! Get us six big glasses of cream soda," ordered Jack.

They drank it all.

"So why wait till after lunch?" asked Jade.

"Because if you fight too early, you might get tired faster." Told Lillith.

"Oh, ok, yea, lunch time is a good time," agreed Jade.

"I think every one of us is ready for this fight, just like you two." Selene pointed it out to Lillith.

"So remember, two of you go to the left, two to the right, then Andrew and I in the middle." Lillith reminded.

"Right!" They all shouted.

"I think we should go spy on the evil, shouldn't we?" Jack explained

"Actually, that sounds like a good idea," admitted Jade.

"Yea, let's go check them out!" Shouted Selene.

“Fine, let’s go.” Lillith smiled, checking out the enemy before attacking them. Good idea.

****

They flew to the east, to where the wendigos seemed to come from. They flew, Lillith took out the blue crystal, and waited for green to show up. It took an hour, but they finally reached the spot where at least two hundred green dots showed up. Lillith raised her hand and yelled. “They are right below, let’s fly back half a mile!” Ordered Lillith.

They landed on the dead earth. As soon as they landed, the dragons started acting sick.

“Thank the Gods I brought the nuts!” Lillith said. She gave each dragon a nut, and they started feeling better, but they all could tell they needed to give their partners a break, so they sat down and relaxed. The dragons now shrank and napped on their partners’ shoulders. They all knew this was going to be challenging, so they pet their partners while they slept. The group all sat and relaxed. They were preparing for this big battle. Were they ready?

“Okay, guys, I want you to fight them at first until we get too tired, if we get tired, but by sword, not magic. Our partners will slice off the giants’ heads with their magic, so they don’t have to get close. Then they won’t get hurt; they are more likely to get hurt.

“We have magic shields, Lillith.” Informed Selene.

“Oh yeah! I forgot.” Lillith laughed.

“Oh, don’t worry about our partners, I know they are listening, they all are, and understand to use long-distance attacks.”

“There are so many,” Dom admitted. Big and regular-sized ones, just wandering around like they are zombies in a jail.”

“Yea, it’s pretty strange.” Selene agreed.

So if I saw correctly, there are spots close to the skinwalker that we can prepare ourselves. I will raise my sceptre, then shoot out a light above the wendigos; that’s when we attack.

Lillith yawned. “Oh no, I’m tired....”

“It’s from all this excitement, let’s just take a nap, I will stay up and watch out, this is a huge battle, a war on evil. You need a nap,” said Jack.

“But, but....” Lillith started to make an excuse.

“Let’s do it, take a short nap, we will wake up even more ready to fight,” said Andrew.

“Okay, let’s nap, we need to relax anyway,” Lillith said.

“Yeah, I’m very stressed out about this, but pumped at the same time,” Selene admitted.

They all lay down except for Jack.

****

Lillith opened her eyes and realised she had fallen asleep right away. She looked up at Andrew, who was just opening up his eyes. They all woke up, except for Jack, who was watching out for them. They all got up and stretched.

It was lunchtime and they were all hungry.

“You guys, if you are anything like me, you are hungry. I brought beef sandwiches, milk, and some beef jerky.” Jack announced.

“FOOD!!!” Dom jokes.

They all ate happily, and their partners loved the beef and bread. And then they all drank the milk.

“Thank the Gods you brought food!” Praised Lillith.

They all ate and drank and felt really ready to fight.

****

“Okay, you know where to go, so we surround them and attack.” Lillith reminded. “Oh, and leave your dragons out of the way, they can attack long distance, and we don’t want our partners hurt.”

“Sounds like a plan,” said Selene.

They all started sneaking up on the group, which took them a good fifteen minutes. They all left their partners, telling them to wait until they started to attack. Lillith got so close she could hear them breathing. Lillith jumped up and shot out the mystical magic into the air. Then, the dragons grew, and they all started shooting the wendigos. They sliced up the wendigo close by at first. The partners shot out the color of an element that was shared with the partner, the wendigo, which were giants. They burned up fast. All of the wendigos started attacking, so they all put up their shields and attacked the wendigo with their weapons, slicing them up, mostly just slicing their heads off. You'd almost feel bad for them if they weren't pure evil. The numbers were going down rapidly. Lillith took out the crystal and looked around. Then, right at the edge, there was a red dot. Lillith dodged and ran to the spot where the red dot was and attacked a hooded person, then stopped. The skinwalker had an infant, Lilith grimaced.

"You coward!"

"Coward, you say?" Says the cloaked skinwalker.

"Yes, hiding behind a baby is very cowardly." Lillith snapped at the skinwalker

"Well, it was just lying there, its parents are long gone, so what should I do with it?" Skinwalker said with a toothy smile..

"Give me the baby, and I won't kill you."

"Now see, was that too hard?" Laughed the skinwalker in its roar of laughter.

It handed the baby to her, and she took it and ran.

"I said I wouldn't kill you, but my partner will! PRAISIA ATTACK THE SKINWALKER BEHIND ME!"

Praisia snorted and blew a purple gust of fire at the skinwalker.

It screams and burns alive until it is nothing but ashes.

"Looks like we are just about done! They had fewer than ten Wendigos left. They all fought the rest and killed them all. One down, four to go." They all laughed.

The baby coed in a strange way, Lillith removed the cover, and the baby turned out to be a severed head of a wendigo.

"Ewww!" Lillith screamed, then threw the head away, then smashed it with her sceptre.

They all headed to Lillith's hometown. It had taken longer than they expected to kill, and they were very tired. So they went to the inn and slept.

****

When Lillith woke up, it was morning, and Andrew was still sleeping. She moved closer to Andrew and put her mouth to his ear and whispered to him.

"I love you so much, my strong, handsome man. You should wake up and give me a big kiss." She giggled. Bob slowly woke up, and Lillith tried to stay still, but he was a smart dragon and tackled Andrew's head, then kissed his face.

"What the? Oh, you little trouble maker," he laughed and sat up, stretching. Lillith wrapped her arms around Andrew's waist.

"Lillith!" He laughed and hugged her back. "We did it!"

"Yes, we did." Lillith purred. "We still have four more to go, then the leader of the skinwalkers.

Lillith felt so glad they did so well.

They took a shower and set off to the pub. Lillith picked up Praisia and thanked her for all her hard work, then thanked Bob.

They entered the pub, and two men were talking, saying three-fourths of the world was dead, seventy-five percent of the people were saved, but how can humanity survive like that? The planet is dead for the most part.

CHAPTER 9 – THE WITCH OF THE WOODS

CHAPTER 9

THE WITCH OF THE WOODS

Lillith's gut sank. She looked at Andrew and whispered, "Three-fourths of the world is dead? Seventy-five percent alive? Are we failing?" Lillith cried.

Andrew held her. "There's got to be a way to save this world, don't worry. We are going to save the world!" Andrew cheered quietly.

Lillith looked up and smiled. "I hope we do." Lillith wiped her eyes dry and straightened up. Praisia licked Lillith's cheek. "It's okay, Praisia, I am just a little sad." Lillith hugged Praisia.

Dom and Jack entered the pub and smiled at the two already there.

"Hey, guys!" Lillith said she couldn't let her team see her sadness because she was hiding it.

"What's up, guys?" Asked Dom.

"Nothing really yet, but I want to go to the Elder witch to ask about the enemy. Sound good?" asked Lillith.

“Sounds good to me,” Jack answered.

“Okay, good.” Lillith smiled.

They all sat at the corner table and waited for the others. Jade and Selene entered, stretching and yawning.

“Good morning, guys,” Selene said to everyone in their group.

The two joined the others at the table.

“Guys, I have a few questions I wanna ask the Elder witch,” Lillith announced.

“Okay, after breakfast.” Agreed Jade.

****

Lillith knocked on the Elder witch’s hut. The Elder witch opened her door, smiling.

“What is going on with you guys? I can help with whatever is needed.”

“Well, we have some questions,” Lillith asked.

“Go ahead and ask.” The Elder witch answered.

“What are Skinwalkers?” Lillith questioned.

“Oh, they once were wise and caring people, but they weren’t happy with the little power they had, so they performed an evil that is helping them kill the world; it transformed them from human to what they are today. The evil gave them this power if they promised to kill the world, and they sealed the pact with blood. Now that they have the dark crystal, they are able to make the pact happen.” The Elder witch answered.

“So they were human?” Asked Jade.

“Yes, they were once witches, but not as strong as me. Few witches are as strong as I; my powers come from a long line of Head witches in my family.”

“Power-hungry monsters!” Jack spat at the thought of selling one’s own soul for power.

“What are wendigos?” Lillith continued to ask, with wide eyes. “Were they humans too?”

“Oh no no no, they are pure evil creatures that survive on energy, such as the energy from the dark crystal and of course the humans they devour.” The Elder witch frowned.

“So, they all must be destroyed, even the Skinwalkers, even though they are all humans?” Lillith asked.

“They are no longer humans anymore, darling. There is no going back for them; they are pure evil.” The Elder witch looked into Lillith’s eyes. “Something is bothering you. What might that be?”

“Today I heard that seventy-five percent of the world is dead. How do we recover? How do we survive on this planet? If you can’t grow or maintain life in the world?”

“Don’t worry about that right now. Know if there’s a will, there’s a way, so don’t let that worry you. Do I look worried? If I looked worried, then you should, but I know you defeated one camp of those creatures. I know you can do it all, hun, once you defeat the last four camps, come home and come to me, we have to make the light crystal by combining all the elements, then you can defeat the head skinwalker who has the dark crystal.”

“We have a lot to do,” Jack replied.

“Yea, but it’s worth it!” Andrew says with enthusiasm, raising a fist as he speaks.

Lillith smiles and hugs Andrew. “You are always positive!”

The Elder witch smiles at the group. “Well, get going, I heard there is a camp near the coast, a couple of hours away.”

“Yea let’s go!” Jade smacked Jack’s back.

“Why did the camps stop attacking?” Lillith asked.

“Because of you guys, you are scaring them!” The Elder witch laughed.

“Good, let them be scared so we can destroy them all!” Selene cheered.

"Thank you," Lillith said and smiled at the Elder witch.

The Elder witch smiled and waved.

****

By the time they reached the camp of wendigos, Lillith noticed they all had a dark purple aura around them that the others couldn't see, but also noticed there were many fewer; they landed behind a small forest.

"Guys, there is something different about these wendigos; they all have a dark purple aura, so please don't forget to put up your shields, just in case, that goes for you too, partners!" All the dragons snorted in agreement to Lillith's request.

"But they are all regular size, I think it should be easier, right?" Asked Dom.

"I don't know, but just be prepared," Lillith warned. I also noticed that the skinwalker is at the very back, brewing something, so stay clear of it. I will take care of it."

"Are you sure you don't need help?" Andrew asked nervously.

"I'm sure, I have Praisia with me for backup." Lillith patted her partner, who was still giant. Praisia rubbed her head on Lillith lightly. Lillith smiled at her. "Ready, guys?"

"LET'S GO KICK SOME EVIL ASS!!" Roared Dom, the others joined in, and they all took out their weapons and ran through the woods to attack the enemy head-on. When they reached the group of wendigos, the wendigos looked up and went crazy, starting to attack. Lillith put up her shield, as did everyone, and her partner did as well. Lillith shot a blade of energy at the first wendigo, but it bounced off.

"Whoa, guys, looks like we are going to have to fight them without magic. I'm going to head to the Skinwalker and see if I can get it and see if the thing that the Skinwalker is making is what's putting up this barrier." Lillith yelled at her team. They all agreed, Lillith jumped on Praisia and climbed up to sit. "Let's go, Praisia!" Praisia roared, which shocked the wendigos, but they only stopped for a moment, then continued to attack. Lillith flew past the group and landed a yard away from the Skinwalker.

"So you came after all, you know, you really are pissing us off." The skinwalker hissed at Lillith.

"Oh, so sorry for upsetting you, for killing innocent people, and our world!" Lillith spat back. "Praisia, stay back, okay, unless you see I need help." Praisia snorted. She didn't like the idea. "I know, but it could hurt you if I'm not careful. I love you, thank you," Lillith took out her sceptre and ran at the skinwalker.

The skinwalker raised a hand, a glob of whatever it was, rose above its head, and then flung it at Lillith. Lillith dodged it, and where the glob hit, it fizzled, as if it were acidic, but more putrid.

"You think that's going to stop me? Nice try, but I'm a lot tougher." She took her sceptre and shot a sword of energy at the skinwalker. The skinwalker tried to dodge, but missed, and it hit its arm.

“Ugh, I will gut you for that!” Screamed the Skinwalker in pain.

“I don’t think you will have time for that!” Lillith yelled at it.

“Do you not see your friends out there struggling? You think you have a chance?”

Lillith looked at her friends. They were struggling. She had better hurry up. She thought. Lillith raised her sceptre and made a scythe shape on top of the sceptre and screamed while attacking.

The skinwalker jumped into the huge pot. Lillith was shocked, but approached slowly. The skinwalker climbed out; it was now twice the size and had a purple aura.

“Now you are helpless to my power!” The skinwalker roared.

Lillith had to dump out the mix to see if that would stop the wendigos from being so formidable. She ran, pretending to run at the skinwalker, then shot an energy blast at the huge pot. It shattered, and the mix flew everywhere, a little bit hit Lillith on the left arm, and boy, did it burn.

“Ha, stupid girl, now you are poisoned! You will die soon enough, a slow and painful death. Lillith didn’t pay it any attention. She looked at the wendigo, and its purple aura was gone. She smiled, gritting her teeth in pain, but she didn’t have time for pain; she ran to the skinwalker and stabbed it in the heart with her crystal sceptre. The skinwalker hissed and punched her away, hissing more. And falling over. “How could you beat me..... how could a human beat me?” It fizzled and turned to dust.

Lillith fell to the ground, dropping her sceptre, and rolled a few feet. Praisia roared and ran to her. She picked up the sceptre and placed it beside Lillith, and then covered her with her wings.

The others were now flying through the crowd of wendigos. Praisia roared at the others, trying to get their attention. Jack was closest and heard. He let the others know something was wrong. Andrew fought hard until he could get through and rushed to Praisia.

"What is wrong, Praisia?" Andrew cried out. Praisia lifted her wings, and there on the ground lay an unconscious Lillith. Her skin was paler than usual.

"We have to get her to the Elder witch! As soon as possible! "Hurry up, guys, we have an emergency!" Andrew lifted Lillith off the ground and saw the burn on her left arm. "Oh no, Lillith, what happened to you?"

Jack ran up to Andrew and looked worried. "What happened?"

"I don't know, but I'm going to take her home to the elder witch. You guys finish up here, okay?" Andrew was trying not to panic, but his words came out shaky.

"Okay, we will, we are almost done anyway, we will meet you at the pub, okay?" Jack replied. "Don't worry, Lillith is tough." Jack tried to comfort Andrew.

"Okay, I know, we are off!" Andrew climbed Praisia with Lillith and her sceptre. Praisia flew off in the direction of their home, and Bob followed. "Hang in there, baby girl." Andrew whispered to Lillith.

****

They landed in front of the Elder witch's hut. Andrew jumped off and ran to the door, banging on it. The Elder witch opened, with a smile on her face, then saw Lillith, looking grey.

"What happened?" Asked the Elder witch.

"I have no idea." Said Andrew.

"Come in, come in, lay her down on the table. quickly, please."

Andrew did as he was told and carefully laid Lillith on the table.

The Elder witch noticed the burn on Lillith's left arm. "Oh no."

"What happened?" Andrew asked, almost yelling.

"She has been poisoned, but not a regular poisoning; this had dark crystal energy in it. Oh my, poor girl, she is suffering."

"How do you know?" Andrew questioned the Elder witch

Because it's negative energy, even if she is passed out, she is suffering. But my God, is she strong. If this were to hit any of you, you'd be dead in seconds."

Andrew held Lillith's hand. It was cool to the touch. "Please tell me she is going to live!" Andrew begged, tears forming in his eyes.

"I hope to the Gods she does, because the world won't last if she dies, we will all be wiped out, and evil will prevail."

"Oh no, Lillith, beat this!" Andrew pleaded.

The Elder witch hurried to the back, grabbed three different pots, and ran back.

"She cleansed the wound, then placed the three lotions on the burn, then wrapped it up. "This is up to Lillith to beat the negative energy. I will give you these lotions. The poison can be cured; it's the energy that needs to be fought. She needs your support. Talk to her, comfort her, just be with her. Feed her this liquid three times a day, and reapply all three lotions five times a day. I put a bandage on her, but let it breathe." The elder witch told Andrew and looked very worried.

"I will," Andrew said.

"Where can I find you? Where are you staying, so I can come check on her?"

"At the inn, room 8," Andrew answered.

"Okay, take her and please, take care of her, she needs you and your love." The Elder witch patted Andrew on the back."

"I will, I promise!" Andrew left carrying Lillith. Praisia saw Andrew, shrank, and flew to Lillith, rubbing her head on Lillith's face. Bob landed on Andrew's shoulder and cooed sadly.

Andrew walked to the inn and entered carefully, not to move Lillith too much, then laid her down on the bed. He held her hand and kissed it. “Come on, Lillith, you can beat this! You are stronger than anyone! I love you...” He began to cry. If only she didn’t face the skinwalker alone.

****

Andrew jumped as a soft knock on the door rang through the small room. Both Praisia and Bob flew at the door protectively.

Bob, Praisia, it’s okay.” Andrew placed Lillith’s hand down and went to the door and answered all his friends from the battlefield. “Hey....”

“How is Lillith?” Asked Jade worried.

“She was poisoned, but it’s not the poison; it was filled with energy from the dark crystal. She beat the poison, but now has to fight out the negative energy.” Andrew said, scratching his head.

“Can we see her?” Asked jack.

“Of course, come in.”

“We thought Lillith would be healed and come to the pub, but after a few hours, we couldn’t wait, and the Elder witch told us you were here; she wouldn’t tell us anything.” Selene cried, and she wiped her eyes.

They all surrounded the bed.

“Did you say negative energy, Andrew?” Jack questioned Andrew.

“Yes, it’s coursing through her; she has to beat it,” Andrew replied.

“Well, I could surround her with positive energy, and it will soak into her until it runs out, plus I have some stones for positive energy, and make them focus on Lillith.” Jack fumbled around. “Yea, I have them here in my bag, they are charged too, I haven’t had to use them since they got charged.”

“I would love all that!” Andrew smiled, “How come the Elder witch didn’t do that?” Andrew wondered.

“Probably because I have to use a person to get the energy from, and that person has to be close to Lillith’s heart, and it could put you in a coma for one to ten days.”

“I will do it!” Andrew announced. “I’m the closest to Lillith.”

Okay, I will be back in a few hours, gotta get supplies.” Jack smiled, then left.

“I’m gonna go grab you a steak and baked potato with lots of butter and sour cream with a jug of cream soda! You need the energy!” Jade announced and ran out the door.

Andrew smiled. “I will force myself to eat, for my love.” He looked down at Lillith, placing his hand on her forehead.

“Well, we will go to the pub and talk about what to do next.” Said Dom

"Yea, we will come back tomorrow." Selene felt Andrew needed to be alone right now.

"Okay, guys, thanks."

They all left and closed the door behind them. Praisia landed on Lillith's shoulder protectively, and Bob landed beside Praisia. Andrew sat beside Lillith. Soon, a knock on the door. Andrew called out to enter.

"I got your food! They threw in peas, hope you like peas!" Jade laughed. She placed the tray on the table. "They also have some meat for Praisia and Bob, and some milk for them too!" The two dragons coed, while Magna coed back at them. "All the dragons are worried," Jade said while patting Magna's head. "Well, I'm out of here. I'm sure Jack should be back soon, so eat up while it's hot!" Jade hugged Andrew, then ran out of the room.

Andrew went to the steak; it smelled so good, and as Jade said, the baked potato was buttery with loads of sour cream. He placed the pieces of meat on the table with the bowl of milk with it.

"Eat up, guys! You need your strength too!" Andrew said calmly. Praisia left Lillith to eat and drink, and Bob followed his mom. Andrew ate everything up, drank most of the cream soda, and went to the bathroom. As he exited the bathroom, he heard a knock at the door. He went to the door. It was Jack, out of breath.

"I got everything, now this magic isn't highly looked upon, it's not black magic, but grey." Jack took out a big satchel. "Go lie next to Lillith, and hold her hand."

"Okay, I'm going to bind your hands together." Informs Jack.

"Wait, if I go into a coma, you have to take the bandage off of Lillith's arm and put these three lotions on her wound at least 5 times a day, but you don't have to put the bandage back on, can I trust you to do that?" Asked Andrew

"Of course I will, I will do it every couple of hours, I'm a witch, so I know what to do. You can trust me." Jack replied.

"Thank you." Andrew lay back down. "And I trust you, Jack."

Jack took out some string and tied their hands together intricately. He took out a dagger. "This is going to hurt a bit; it needs blood to perform. Just a little cut on your other hand and a little cut on Lillith's."

"Okay." Andrew closed his eyes.

Jack took the dagger and cut Lillith's hand, then moved over to Andrew and cut his hand. He placed herbs and honey on both wounds. Then sat at the base with his hands and began to manifest the spell. It took half an hour to get it all out. He then waved his wand, and a white light surrounded Lillith. He checked their hearts; they were beating at the same pace. It was working. He placed the three quartz crystals around Lillith, which made the white light glow stronger. He then placed a rose quartz crystal between the two, which made the glow stronger, and a faint pink light glowed around Andrew. The white light was feeding into Lillith, and the pink light was feeding into Andrew.

"Okay, done." He pondered if he should put the lotions on Lillith? He decided to take off the bandage. The burn looked bad. He rubbed lotion on the burn until it was soaked. He prayed for them both. Andrew was out cold, but he knew it was helping. Lillith went from a soft grey to a pale skin color. He prayed to the Gods that this work. Jack was very tired after all the work he had done.

****

Jack went into the pub, and everyone was there and looked at Jack.

"How did it go?" Asked Jade.

I think it went well, Lillith isn't looking grey anymore. But the two are out of commission for a few days. I did everything I knew to get Lillith better, and now Andrew is in a coma, but only for a short time. I have to go back every few hours, but look at who I brought." Jack smiled as he now had three dragons on his shoulders.

"I have an idea, why not all of us, not including you, Jack, but we can all split up looking for where the last three spots are where the wendigos are located, so we don't waste time," Dom suggested.

"That is an amazing idea! We can check out where and what the fight might be like." Jade said excitedly.

Okay, so we have a plan. We will start tomorrow. I don't know about you guys, but I'm exhausted. I'm going to go to Lillith and Andrew's room and sleep." Jack said while stretching.

"Sounds good to me," Dom admitted. They all left for the inn.

****

While the others searched the world for the camps of wendigos, Jack took care of Lillith. On the third day, Andrew woke up.

“Ugh, how long was I out for?” Andrew asked Jack.

“Three days, but everything is going great, Lillith even has some rosiness in her cheeks, look!” Jack said happily.

Andrew looked at Lillith, and she was still out, but she looked beautiful. Andrew winced at the pain in his hand from the cut. Just then, someone knocked at the door, and Jack answered it. It was the Elder witch she asked to come in, and she was welcomed in. Andrew sat up and smiled at her.

The Elder witch looked at the two on the bed. “Jack, did you use grey magic?”

Jack looked down. “I know it was wrong, but it felt right at the time. Please don’t tell anyone.” Jack pleaded.

The Elder witch walked over to Lillith and put her hand on Lillith’s forehead. “Hmmm.”

Jack stood there ready to be yelled at.

“She looks much better, thank you, Jack, for your work. I was so scared she wasn’t going to make it! But the two of you brought her back, “ The Elder witch hugged Jack. “You saved us all!” She began to cry.

Jack hugged her back and let her cry. The Elder witch went to see the burn, and it looked great; there would be a scar, but nothing too bad.

"I'm going to unbind you two now, Lillith looks like she's just sleeping." Jack unbound the hands.

"I'm going to try and wake her." Announced the Elder witch. She walked over and patted Lillith's cheek. "Come on, sweetheart, wake up, you can beat all evil! You are our saviour."

Lillith stirred, but didn't wake up. Praisia flew to Lillith and licked her face. Lillith slowly stirred, then opened up her eyes.

"Praisia! I remember fighting. I don't remember anything after that. What happened?" Lillith asked.

"You just about died, but because of these two men, you survived."

Lillith looked shocked. "Okay, Praisia and Andrew, you and Bob are going to help me fight the next three skinwalkers! I can't die!" She began to cry.

Andrew held her tight, then kissed her forehead.

"Where is everyone?" Lillith asked.

"They are searching for the last three camps; we found two so far. Hopefully, we will find the third today." Told Jack.

“Oh, that would be great!” Lillith grew excited, but fell back onto the bed. “I feel so... I don’t know, I’m very weak.”

“You beat something that was supposed to kill you; your body needs rest.” The elder witch explained.

“It was that bad?” Lillith was shocked.

“Yes, and I think you should stay in bed for a couple of days,” Jack told Lillith.

“But I’m starving! Andrew, can you help me get to the pub?”

“No! You rest.”

“I don’t want to rest, the more I move, the faster I get better!” Lillith insisted.

The Elder witch looked at Lillith. “I think she’s right, the more she moves, the faster she will heal.”

“WHAT?” Both Jack and Andrew exclaimed.

Lillith stood up; she was wobbly.” Andrew, help me.”

“Okay, if it will help you,” Andrew said and helped Lillith steady herself.

The Elder witch said bye, and they were off to the pub.

****

Lillith wobbled into the pub and sat in the corner booth. “I wonder when the others will return?”

“They are working extra hard for you, Lillith,” Jack told her.

“So you guys know where the two camps are? Anything like the last one?” Lillith asked.

“One has all the giants, the other is like the wendigos are on fire. But they aren’t burning, it’s strange.” Jack told them.

“Well, they better be prepared, we are ready for them.” Lillith laughed, then ordered a burger with her now favourite drink, cream soda.

“Andrew held Lillith’s hand. “We need you to heal fully.”

“Oh crap, I have a question for the Elder witch. I had my shield up; how could the goop hit me?”

“The Elder witch told me it had the dark crystal energy, maybe that is why?” Andrew answered.

“That makes sense.” Lillith nodded. “Yea, that is the best answer.” She ate her burger and cream soda and gave some of the burger and lettuce to Praisia.

They all let their dragons onto the table after they were all done. They played around, nudging each other and pouncing on each other, having a blast.

Lillith leaned on Andrew. "Aren't they precious?"

Jack laughed. "Yes, they are."

Later, Jade joined. "Hey, OH MY GODS! LILLITH!!" She hugged Lillith as if they hadn't seen each other in years. "We were all so worried about you! Oh, that burn looks like it hurts."

"I'm good with pain, always have been," Lillith admitted.

"That is very true." Andrew smiled at Jade.

"I didn't find the last camp, but I'm sure one of the others has. We worked hard for you, Lillith." Jade laughed.

"Yes, you guys did, you were very productive, I'm proud of you guys." Lillith complimented her and the others who hadn't returned.

"We did all we could," Jade said with a smile on her face.

Dom and Selene entered, and when they saw Lillith, they cheered and hugged her.

"Any luck, guys?" Asked Lillith.

Dom smiled, found the last one. It's filled with wendigos, but it's strange; they are all shadows, like ghosts."

"That is strange, but I'm sure our magic will work on them," Lillith told.

"Yea, no doubt." Dom smiled. Dom and Seline ordered their dinners and ate.

"So I was thinking, instead of just me going for the skinwalker, Andrew and our two dragons will fight them, while you guys fight and destroy the wendigos, sound good?" Asked Lillith.

"You don't need to ask us for permission; you are the leader, what you say goes," Jack said.

"I need to build up my energy, so, you guys, I want you to practise fighting and using your powers, you too, partners." The dragons coed in agreement.

"What time is it, Jack?" Asked Lillith.

"It's 12:30, getting late," Jack said.

"Okay, guys, let's get to bed. Andrew and I will be working on me to get my strength back as you guys are working," Lillith informed.

"Sounds great!" Dom replied.

They all headed to the inn. Lillith was still wobbly, but she tried her hardest. Nothing was going to stop her from saving this world!

****

When Lillith got into bed, she pulled Andrew on top of her, then kissed him deeply.

“Oh, how I missed you.” Uttered Andrew. “I never left your side.”

“And you never will.” Lillith laughed and held Andrew close, squishing him a little, because she didn’t have all her strength.”

“You are an angel, my beautiful soon-to-be wife.” Andrew purred softly, then continued to make out with her. They hadn’t been this close in so long. “This feels so natural and good.”

They lay down on each side of the bed and held each other until they both fell asleep in each other’s arms.

CHAPTER 10 – SILVER FIRE

CHAPTER 10

SILVER FIRE

Lillith woke up in Andrew's arms; it was so nice and warm.

"Good morning," Andrew whispered to Lillith.

Lillith looked into Andrew's eyes. "You dare wake up before me?" She laughed.

"We have a lot of work today, are you ready?" Asked Andrew.

"Am I not always ready?" Lillith jokes.

Andrew laughed. "True, how does your burn feel?"

"It burns, but I can handle it." She looked at the burn on her left arm. "I think we should wrap it up today."

"I will then, but first let's get you in the shower. You need help in the shower?" Andrew asked.

"I should be fine. If you hear a thud, I just fell, I will leave the door unlocked for just an emergency." She smiled.

"Okay, hun, get your butt up and into the shower!" Andrew ordered playfully.

Just then, they were attacked with love from Praisia and Bob.

"Good morning, guys!" Lillith laughed and snuggled Praisia, who loved it. "Okay, Praisia, it's time for a shower. Are you ready?"

Praisia coed and rubbed against Lillith's head.

Lillith tried to get up, but it took a few tries, and she finally managed to stand up. Andrew frowned. He wanted to help Lillith, but he knew she wouldn't like that.

Lillith wobbled to the bathroom. "Be back in a few minutes, hun." She told Andrew.

"Okay, I will be right here," Andrew confirmed.

After the two took showers, they decided to go to breakfast. So they walked to the pub, Lillith wobbling, but refused help, which honestly made Andrew sad, but also proud of Lillith.

They ordered two eggs and a few sausages. They both gave Praisia and Bob a sausage, along with some milk and water. Lillith and Andrew had apple juice. They were going to wait for everyone, but Lillith really wanted to train her body. So they went for a walk, and Lillith was annoyed that her legs were not working properly.

“Andrew, let’s practise fighting with swords,” Lillith said excitedly.

“Lillith! Don’t push yourself too hard!” Warned Andrew.

“I know what I can do, and I need to get my strength back in my arms. Just think, you can actually beat me now.” She laughed and stuck out her tongue at Andrew.

“Okay, but they will be wooden swords, please?” Andrew begged.

Lillith sighed. “Fine, but we have to train just like everyone else. I understand you will understand when I need to take a break.” Lillith sighed.

“Of course, hun, I love you and I don’t care if we have to take breaks every five minutes, as long as you are okay,” Andrew replied.

“You are the best, trust me, it will be worth it,” Lillith told.

Lillith was already getting better at walking; she wasn’t wobbling as much, which made her very happy.

They visited where her father trained others, which was a high-demand course to take. Lillith went to grab a wooden sword. She couldn’t believe how hard it was to pick up, but she managed. “Are you ready to kick my butt?” Lillith laughed.

“We are not fighting until you can use the sword, so follow me to build up your strength,” Andrew explained.

“Okay, that’s fair.” Lillith agreed.

Lillith lifted the wooden sword and tried to strike at nothing. She almost fell over, but that didn’t stop her. She practised until lunchtime. Lillith was exhausted, but she felt good. She loved every second of it, even if she almost fell over numerous times, but she was getting stronger. Lillith was weak now from all the work she had done. She was regaining her balance, and her legs were moving faster. Her arms were sore, but that just meant the training was working. It was lunchtime, and they were starving.

“Let’s take a break and get some lunch?” Andrew suggested.

“Sounds great to me.” She smiled and kissed Andrew on the cheek. “Thank you for helping me. It means the world to me.” Lillith smiled at Andrew.

“Of course I’m going to help you, you are the love of my life, I would do anything for you.” Andrew hugged Lillith. “Are you ready to go?”

Lillith walked, much better than before, but her legs were still shaky. She placed the wooden sword back and started for the door. Andrew followed, ready to catch Lillith, but he doubted she would need help; she was so strong. Lillith walked without help to the pub; everyone was there.

“Hey, guys!” Lillith said cheerfully.

“Hey sunshine!” Dom said. “Looks like you have been working hard, you are barely wobbling. What have you done today so far?”

Lillith smiled. “I’m practising fighting with Andrew, I’m using a wooden sword, but it still feels heavy, but I’m getting stronger and more in control,” Lillith said proudly.

“I swear, Lillith is the strongest person I know, and she is just so amazing.” Andrew bragged.

“You two are so in love, you two are such a cute couple!” Selene praised, then Selene smiled and held Lillith’s hand.

“You got this!” Jade encouraged Lillith.

“Did you guys already eat?” Asked Andrew.

“Nope, we were hoping you two would show up,” Dom said.

“Aww, you guys are the best. How’s the training going?” Asked Lillith.

“It is going great, we are sparing, not using magic yet, we will do that after lunch,” Jack informed them.

That’s amazing, you guys are great team mates!” Lillith almost yelled in excitement. “Oh, are your partners training too?” Asked Lillith.

“Oh yes, they are, Dom keeps making Ice in big chunks, and they shoot at them,” said Jade.

Lillith looked at Praisia, then at Bob. “Can you help train these two? I want them to be as strong as possible." Lillith petted Praisia, but Praisia was not happy. "Come on, you two, the fights are getting stronger, you need to practise," Lillith explained. Praisia hopped off Lillith and flew up on Dom's shoulder. Bob refused.

"Bob, you have to get stronger, please, just go along with it." Andrew pleaded.

Bob jumped off and snorted. He always wanted to be with Andrew, but he had to listen to avoid upsetting Andrew. He flew onto Jack's shoulder and rubbed against Merlo. Praisia looked at her son, then squealed at his stubbornness. Bob squeaked back and lay down.

They all had the special of the day, nice and tasty BLTs. They all gave their partners a piece of bacon and made sure they drank too.

Lillith finished her cream soda and looked at all her friends. "Let's get back to work!" Lillith said excitedly.

"Lillith, relax for at least a little longer, you are working yourself too hard, if you push yourself too hard, you will burn out." Warned Jack.

"I feel great. I was tired when I got here, but now I feel rested and ready," Lillith told everyone.

"There's no use, when Lillith gets into training, she could do it for hours, which she has done, her willpower is something so powerful. So I think we should do what she says, she knows her body and knows when to take breaks." Andrew explained.

“Okay then, but please rest once in a while,” Selene told Lillith.

“I will, promise.” Lillith smiled, slowly getting up. Praisia squeaked. She was going to miss Lillith.

Andrew followed. They all noticed how much better Lillith was already. She barely wobbled.

****

Lillith thrusted, struck a wooden pole, and continued for hours. She could feel her body getting stronger. Until she accidentally broke the wooden sword.

“Crap!” Lillith felt bad for breaking one of her father’s training swords.

“Lillith, I think it’s time to work with a metal sword. You are doing so well,” Andrew told her.

“I’m getting stronger every hour!” Lillith cheered. Andrew cleaned up the mess of the broken wooden sword.

“You know you are an angel, my love.” Said Lillith to her future husband. She began to walk more steadfastly, picking up a metal sword. “Ahhhh, this feels good!” Lillith swung it around. It was hard, but if she practised, she would regain her strength and be able to fight.

Out of nowhere, the two heard people screaming.

Lillith looked at Andrew and ran as well as she could towards the screaming. Then, when they got outside, they saw five giant wendigos coming from the forest. Jade, Jack, Dom, and Seline met with Lillith.

“What is going on?”

“Lillith, you stay out of this. We can take care of this.” Andrew ordered Lillith.

“Fine....” Lillith felt useless.

Jade and Jack attacked the first wendigo; they sliced its head off. The strange thing is, the wendigos had no protection spell on them, so it was easy to take them all out.”

Lillith sighed; she hated being useless.

“We need to have a meeting, let’s go to the pub, the villagers took care of the bodies.

****

Lillith looked miserable; everyone looked at her. Andrew put his arm around Lillith to comfort her.

“Okay, guys.” Jack started. “I think they think you didn’t survive, so they are testing us. Little do they know Lillith is one strong woman!”

“Yea, that was a weak attack, no one got hurt, and no houses were damaged,” Andrew said.

We have to destroy these camps, so now we have to work harder and faster, got that, guys?" Lillith decided.

"Okay, well, you guys go train yourselves and the dragons. Lillith and I will continue, we are doing amazing. The skinwalkers are scared of us; they just wanted proof you were still alive, Lillith." Andrew told them.

"They are going to think I'm dead because I didn't fight." Lillith worried.

"Good, they can get cocky, and we will destroy them." Dom cheered.

"I think that's an advantage; they think we are weak, little do they know how strong we are getting, especially you, Lillith. You may not know this, but you are emitting a very strong aura of power." Jack told Lillith.

That put a big smile on her face. "Really?" Asked Lillith.

"Oh yes, your aura is so intimidating, I'm just glad you are on our side!" Jack laughed.

"Let's go back to training. I want to get stronger, and you know that saying, what doesn't kill you makes you stronger." Lillith smiled.

"You really are right about that!" Andrew admitted, "You are so strong. You know she broke a wooden sword, now she's using a regular sword." Soon we can practise our magic." Andrew said proudly of his love.

****

Lillith mastered the sword in a day. She walked normally now. On the third day, Lillith and Andrew went to work on the magic attacks. Everyone cheered happily.

****

Lillith sat in the corner booth while the entire team sat with her.

"Okay, guys, I think we are ready to get back into fighting the wendigos and the skinwalkers. Which camp is closest?" Asked Jade.

"The one with all the giant wendigos. We will defeat another skinwalker, but don't worry, Andrew will be helping me, including Praisia and Bob." Lillith said.

"If you feel ready, then we all are ready." Jade agreed.

****

Jade led the way on Magna and landed a field away.

"Okay, they don't have any dark purple auras, just be careful, you don't know what magic the skinwalker is using," Lillith ordered.

Lillith and Andrew climbed back on Praisia and Bob and flew to find the skinwalker. The rest of the team ran towards the giant wendigos and started chopping off their heads, but something strange was happening. Vines grew out of the neck and reattached its head, and it stood up.

"What?" Jack said, surprised. "Okay, guys, we have one heck of a fight ahead of us. Magna shot a blast of fire at one. It burned it, and it fell. But as soon as the fire went out, the charred body got back up.

"This is crazy! It must be a spell. But like last time, Lillith stopped the magic on the last ones. Let's just hope she stops this crap from happening," hoped Selene.

****

Lillith flew around, took out the crystal, and saw a shack with a red dot on it. "It's in the shack, let's go!" Lillith landed feet away from the shack, and Andrew landed behind Lillith.

"Burn the hut down, Praisia!" Lillith ordered.

Praisia blew lilac fire until the entire shack was on fire. The skinwalker ran out, carrying vines with it.

"Y-you are supposed to be dead!" It hissed.

"Sorry, I'm not that easily killed," Lillith said in an angry tone.

The skinwalker laughed at Lillith. "You don't stand a chance against my magic!" He closed his eyes, and the vines came alive and tried to strike Lillith, but Lillith jumped out of the way.

Lillith took her sceptre. "Andrew, get your crystal sword out, we need to trim some plants!" She laughed as Andrew took out his sword.

Lillith shot a powerful shard of energy at the skinwalker. It hit it, and he screamed, as one of his arms fell to the ground. Andrew ran at the vine and chopped off a big chunk. The chunk squirmed on the ground, then poofed into ashes, but the vine was still growing and attacking them. Lillith jumped up and sliced the container holding the vines in half, then threw energy at the vines. It shrivelled up, the skinwalker was shocked, and you could see its fear.

"You guys actually think you can beat us? Good luck!" Lillith yelled at the top of her voice.

The skinwalker screamed in anger at the two.

Praisia, destroy this evil creature!" Lilith ordered.

Praisia snorted, then blew a giant beam of lilac purple energy at the skinwalker.

While the skinwalker began to burn, he screamed. "You will never beat our leader; we are ants compared to them." Then turned to ashes.

As soon as the skinwalker was dead, the wendigo stopped growing back together; there were no more vines. Lillith and Andrew turned to the wendigos and started chopping their heads off, one by one. They killed the last one and cheered.

"Lillith, they kept rejuvenating until you destroyed the skinwalker! And thank the Gods you did, because they were surrounding us." Dom said happily and out of breath.

“Well, it was Praisia that killed him. I think we should treat all our partners to a special dinner at the pub! With milk and honey mixed in.” Lillith laughed. All the dragons roared happily; they deserved a feast.

They all headed back to Lillith’s hometown and went to the pub. It was now turning dark, and they all were tired; it was a long day.

“Three down, two to go,” Jade said happily.

“The last two are going to be hard, I mean, the wendigo on one camp are on fire, and the other one they are shadows. Have no clue how to fight shadows.” Lillith worried.

“We will do it, we will beat them all,” Jack said triumphantly.

They entered the pub and ordered a chopped-up steak, with four bowls of milk with honey mixed in. Praisia squealed in delight, and the other dragons followed suit with Praisia.

They finished everything. The dragons finished everything, too. They ate and drank until nothing was left.

“You guys are probably happy and ready to sleep,” Lillith asked the dragons. They all purred happily and climbed up their partners’ arms and curled up on their shoulders. “Aww, what sweet babies!” Lillith said softly and cuddled Praisia, who licked Lillith’s cheek.

They all decided to go to bed; they were exhausted from the battle they had fought.

****

Lillith curled up next to Andrew on the bed. She began to cry hard.

"What is wrong, Lillith?" Andrew asked, pulling her closer to him, trying to comfort her."

"I just am scared to fight the head skinwalker, I have to fight it alone, and obviously I can't fight alone.....When I did that, I almost died, and if I fail or die, the world is doomed...." Lillith hid her face in Andrew's chest, still crying.

"Hun, it was an accident that you even got hit by that goop; it was a fluke, you are stronger than anyone on this planet. And by then, we will combine all the elements together, making you stronger than those evil creatures." Andrew told her.

"You think so?" Lillith sniffled.

"I know so, you are the best, better than the rest." Andrew kissed Lillith's forehead.

Lillith wiped her eyes and started laughing. "I'm going to be the best wife to you." She looked into Andrew's eyes. "You will be spoiled and will always be loved." Lillith leaned over and kissed Andrew deeply, then they held each other until they fell asleep.

Praisia, who was lying on the chair with her son Bob, cuddled with him and fell asleep.

****

The next morning, Lillith woke up first, as usual, in Andrew's arms. She could feel his heartbeat; it was bliss. Andrew's long blond hair was out of his braid, covering up his face. Lillith moved his hair and whispered to him. "I love you and only you for all time, then kissed him softly on the cheek. They stayed like that until Andrew started to wake up Lillith smiled at him.

"Good morning, my love." Lillith kissed him softly.

"If I woke up to this every morning, I'd die a happy death."

"Don't ever say you are going to die! You are my best friend, my fiancé, and the most treasured person in the world to me. You are living a nice, long life!" Lillith said with love but fear as well.

"Okay, for you I will!" Andrew said calmly.

Lillith smiled and slowly got up.

"I'm going to go jump in the shower; you are after me." She smiled and left the bed.

Andrew lay back down with his arms behind his head and breathed a nice, relaxing breath. He was so happy to have Lillith; she was such a great person, and she loved him. Life couldn't get better; he was truly a happy man.

****

The happy couple entered the pub, and everyone was already in there waiting for them.

"Jack, what time is it?" Asked Lillith, surprised.

"It is twelve thirty. You guys really slept in."

"Ooops," Andrew said, laughing.

"We needed it," Lillith admitted.

"We all just got up too." Dom laughed. "We all slept in." They all laughed.

"Let's eat some breakfast/lunch now, then figure out what we are going to do," Lillith said.

"Sounds good!" Selene agreed.

They all had a burger and some cream soda. Once they finished, they got serious.

"So which camp is closest?" Lillith asked.

"The one with the wendigos on fire," Jade answered.

"Hmmm.... Do you think you can put out the fires with your water, Dom?" Asked Lillith.

“I will try, probably a few at a time, then the others can kill them while I put out the others, TEAM WORK!” Dom cheered.

“Well, I bet they are on fire because of the skinwalker’s power. If I kill it, the fire will go out. Just keep the wendigo’s busy while I kill the skinwalker.” Lillith planned. “Sounds good?”

“We kill the skinwalker.” Andrew pointed out. “And don’t forget Praisia and Bob.”

“Yes! We will.” Lillith said, smiling.

“Okay, we will try everything,” Jack agreed.

“Okay, who knows where that camp is?” Lillith asked.

“I do, I’m the one who found it,” Dom said proudly.

“Perfect, let’s go!” Lillith exclaimed.

They left the pub. The dragons jumped off their partners, flew, and then turned huge. They climbed on and flew off to their destination.”

****

They landed in the field, and boy, did it smell like cooked rotten flesh.

"Ugh, Gawd, the stench is overpowering," Jack complained.

"It is, but we have to destroy evil, so we have to ignore it..... sorry guys," Lillith explained.

"Okay, let's go find it, Andrew," Lillith commanded.

"Yes, my angel." Andrew was ready for action.

The two climbed onto their partners and flew into the air, looking for the skinwalker. Lillith took out the crystal that shows red where the skinwalker would be.

"Bingo!" Lillith pointed to the southern side of the camp of wendigos. They flew down and landed a good distance from the skinwalker.

"So you found me? Good, now I can eat your still beating hearts." It laughed at the two.

"Not before we turn you into ashes!" Yelled Lillith.

The skinwalker raised his hand, and a purple flame appeared. He then blew on it, and a huge flame came crashing down on Lillith and her team. They dodged it.

"Don't let the fire hit you, I bet it's poisonous like the goop was," Lillith yelled at Andrew, Praisia, and Bob. "Barriers don't work against it either." Lillith pointed out. They all circled the Skinwalker, dodging the fire.

****

Jade and the others, including their partners, went to the Wendigos.

"I'm going to try to put them out!" Yelled Dom. He raised his hand, picturing a huge ball of water. The ball started to form until it was huge, then he shot it at the wendigos. It put out the fires, but as soon as they went out, they lit back up.

"This is going to be harder than we thought......" Dom announced, annoyed. They started shooting their powers at them. The flames blocked the bursts of energy.

"We can't fight them," Selene complained.

The wendigos started running towards them.

"Run!" Yelled Jack.

****

Praisia started blasting the skinwalker with her magic flames, but nothing happened. She kept it busy, then Bob blasted it with his power, and Andrew blasted it with his energy. The three kept it busy while Lillith snuck up to the skinwalker and stabbed it in the heart. It screamed then fell.

“You will not kill me....” It said.

Praisia glared at the skinwalker and blew a huge amount of flames at it. It screamed, then in a flash, it was ashes.

****

All the flames disappeared.

Jade yelled, “The flames are gone, let’s kick some butt!” They turned around and started a war cry, chopping their heads off left and right, until there were none left. Then they noticed Lillith and Andrew were killing them, too. They finished them all off within an hour. They were all out of breath.

“We did it!” Cheered Jade and Jack.

“I dunno, is it just me, or are you guys starving and exhausted at the same time?” Asked Dom.

They all agreed, they climbed onto their partners and flew home.

****

They ordered food and ate.

“You guys, I’m too tired to finish,” Jade confessed.

“I think we all are,” Andrew replied.

“Okay, let’s go to bed,” Lillith said.

They all walked to the inn. As soon as Lillith entered the room, she pulled Andrew to the bed. They both flopped on the bed, holding hands, and fell asleep instantly.

****

A dark entity entered the town of Koabi, walking towards Lillith’s parents’ house. It goes to the door, tries to open it, and pulls back. Its hand is burned. It then noticed the charm placed above the door many years ago. “I may not be able to enter, but can you get out?” It laughed as it put a powerful spell on the door and all the windows. It looked through the keyhole. “Perfect!” It raised its hand to the keyhole and blew green smoke into it, filling the home with poison. “There, you are killing my family, so I’m killing your family.” Laughed the skinwalker.

Gwen woke up coughing, and she woke up Luthor. “Something’s wrong!”

Luthor couldn’t stop coughing. “What’s going on? Asked Luthor. “Let’s get out of here!” Yelled Luthor.

They ran to the front door. Luthor tried to open the door, but it wouldn’t budge. Luthor and Gwen couldn’t stop coughing; it was so hard to breathe.

Gwen was getting weak.

“I think we are going to die.” Luthor said sadly, “This is magic, ‘we don’t stand a chance.’”

Gwen looked at her beloved husband. "Let's lie in bed and hold each other, I want to die in your arms...." Gwen said sadly.

Luthor fell to his knees. "Yes, let's do that before we die."

Gwen started crying, they got into bed and held each other, crying, worrying that Lillith would feel responsible for their deaths. They held each other until they stopped breathing. They passed away, crying for Lillith.

****

Lillith woke up out of a deep sleep, tears running down her face. The jolt woke Andrew up.

"What is wrong, Lillith?"

"I don't know, I just feel like something bad happened."

Andrew hugged Lillith and kissed her on her neck. "I'm sure everything is fine." Trying to stay positive.

"Something bad happened, hun." Streams of tears ran down her face.

"It could have been a bad dream, but if it's not, I'm here for you." He squeezed Lillith's hand.

Lillith squeezed back. "I think all of us should see dad and mom. I don't know why I feel this way. Oh Gods, I don't like this feeling. Let's wake up everyone and go check now."

"Hun, wait until morning, you need more sleep." Andrew ran his fingers through her hair.

"Okay.... This is a terrible feeling." Lillith explained.

"Don't worry, all of us have your back if anything should happen." Soothed Andrew.

"I'm grateful I have you and the others," Lillith whispered.

Praisia and Bob flew and landed on their partner's shoulders. Praisia licked the tears away, then nudged Lillith. Lillith pet Praisia and apologized for waking her up. Praisia snorted, then climbed to Lillith's chest, and lay on where Lillith's heart was. Praisia knew something went wrong, so wrong, but she couldn't speak, something even Lillith could feel. Praisia was very worried. She knew this could break Lillith. She was unsure of what had happened, but Praisia knew it was bad, very bad.

"Lillith, please fall asleep. If I stay up all night trying to get you to sleep, that wouldn't bug me one bit. You are the love, my treasure, my beautiful gift from the Gods." Andrew said, then kissed her softly.

"I know, and now I'm grateful, I just have a very bad feeling, I'm scared of what happened, if something happened. But I feel it has to do with my parents."

Your dad could fight anyone off, so don't worry. He is like you, and doesn't give up until the trouble is over." Andrew said softly.

Lillith lay her head on Andrew's shoulder, Bob jumping out of the way and joining his mother on Lillith. Bob could sense something, too. He cooed at his mother. She squawked back and let him come cuddle with her. He wanted to comfort Lillith and his mother.

Praisia got a shocked look on her face, then tears started forming in her eyes. She didn't want Lillith to suffer, but it was too late. She knew now what had happened, but she couldn't tell Lillith because they couldn't speak. All four of them cuddled and tried to comfort Lillith. Lillith fell asleep while crying. Andrew snuggled Lillith. He wished he could stop this feeling in Lillith. He was worried about what might have happened. He was so worried that he stayed up the whole night looking after Lillith.

Chapter 11 – Blades and Beginnings

CHAPTER 11

BLADES AND BEGINNINGS

Lillith woke up sat up, and saw that Andrew was still sleeping. Should she go check on her parents without her team? No, because they are a team, they work together, plus her mom and dad would love a visit from them all. She slowly caresses Andrew's face, God knows you are my one and only."

Andrew slowly woke up and looked into Lillith's eyes. He never saw her so worried. He hoped this was a false alarm; he didn't want Lillith to be sad, not even a second.

"Let's get the others and go to Mom and Dad's," Lillith said urgently.

"Okay." Andrew agreed. They took their showers, got dressed, and then knocked on the ninth room's door.

Jack opened the door, still tired, but then noticed Lillith looking panicked.

"What is wrong, Lillith?" Jack asked, concerned.

"I have a bad feeling, we need to check on my parents."

Jack woke up Dom, and then they knocked on the tenth door of the inn. Jade and Seline opened the door, noticing the worry in Lillith's eyes.

"Oh my Gods, what is wrong, hun?" Asked Selene.

"I don't know, I'm just worried about my parents, can you come with us to see if I'm just being paranoid?" Lillith explained.

"Okay, just need to get dressed, be right with you." They closed the door, and Jade whispered to Seline. "I'm scared, I feel Lillith is right on the problem, and I'm scared for her of what might or might not happen."

"All we can do is support her and protect her."

Jade and Seline finished getting ready and were soon walking with the group to Lillith's house.

Praisia flew off of Lillith and tried stopping Lillith from going.

"Praisia, it's most likely a false alarm; everything's good, I just have the urge to check on them." Lillith had a little smile on her face, but Lillith's eyes told a whole different story. Praisia huffed and landed back on Lillith's shoulder and curled her tail around Lillith's neck.

They walked up to the house, and Jack stopped them. "Something is wrong; there is a circle of negative energy surrounding the house. This spell locked them in," Jack said, alarmed. "I will get rid of it." Jack took out his staff and blew all the energy and dangerous herbs. "Okay, we can enter." Lillith went into the house, then rushed out.

"There's something wrong, something green in the air."

I will get it out." Jack took his staff out and sucked all the gas in one stroke.

"Thank you, Jack. I need to see my mom and dad!" Lillith ran in, and they tried to stop her, but she was too fast. Andrew followed Lillith. The others stayed outside. Jack looked at the others and had a big frown on his face. "That was toxic poison...... I feel sick, poor Lillith."

Lillith ran into her parents' bedroom. "Oh, good, they are just sleeping." She walked to the bed. "Dad, Mom, wake up, sleepy heads." They didn't answer, so she shook them; no response. She realised they were gone. "Wake up, Dad, Mom, please open your eyes, guys, this isn't funny! wake up, OPEN YOUR EYES DAMMIT!" Tears poured down her face. "Come on, Andrew, help me get them to the Elder witch!"

Andrew walked over to the bed and put his fingers to both of their necks; they were both cold to the touch. "Lillith, I'm sorry to say they are gone. Cold to the touch, no heartbeat either."

Lillith screamed a loud anguished "NOOOOOOOOO!" Lillith hugged her parents, but there was no response. "No, no, nooooo!" Please wake up dad, mom, please....."

Andrew went to Lillith's side and hugged her tight.

Lillith cried loudly and out of control, she fell to her knees and covered her face, crying harder than ever.

Andrew felt useless; all he could do was comfort Lillith by being by her side. Tears filled his eyes; he loved Lillith's family, they were people he looked up to and cared about. He was overcome with pain, but he had to be strong for Lillith.

Lillith stood up, wiping her eyes, which were now puffy and red.

"What do we do now?" She then hugged Andrew as hard as she could, crying. "What do we do?"

"We give them a great funeral."

"What??? I can't put them into the ground! It's cold and lonely!" Lillith exclaimed.

"How about we bury them together? My dad can make the casket."

"Lillith hid her face on Andrew's shoulder. "I want them back, I need them, they raised me, they fed and loved me, and I couldn't save them! I failed them! They probably hate me!" She cried harder.

"No, my darling Lillith, they love you, they are here with you, but on another level. I bet you your dad and mom are still together, still loving each other. They love you more than you know. You know how many times they would tell me how proud they were of you. They always said you were their blessing from the Gods." Andrew comforted Lillith.

"Will you help me with the burial, not with money, we have more than enough in the safe."

Lillith kissed both her mother's and father's foreheads, which were cold. I'm going to make the funeral as fancy as possible; they deserve to be sent off with a great send-off. I have to get out of here, just thinking of them in the ground is killing me." Lillith left the house and met Jack, Jade, Dom, and Selene. Lillth wiped her eyes. Lillith told them both that her parents were dead. They all hugged Lillith. Lillith cried again; she hadn't visited her parents in a few weeks, as she was too busy. "Sorry, guys, we have to plan a funeral for both my parents. Do you mind waiting?"

"Take your time, we only have one more camp to destroy, it can wait, they are not killing people or the world. So it's okay."

These skinwalkers better be prepared. We are all going to fight the skinwalkers, all at once." Lillith said, sniffling.

"That is a good plan. The last two are going to be hard, especially the last one. After we make the light crystal, we will still have our powers, we will obliterate their existence! They must be trying to stop you, I mean, this was planned out, they are trying to break you," Selene said softly.

"The wendigo won't even notice until it's over. So we can focus on the skinwalker.

****

Men from the burial service carried out both parents on a board. An arm fell off the board and hung where Lillith could see; she recognised the arm, it was her father's. Lillith bit her lip, then turned, leaning into Andrew. Andrew held her tight. "It's going to be okay, they will always be in our hearts."

Lillith looked into Andrew's eyes, and Andrew could see flames in her eyes.

"I'm going to destroy those bastards! They are going to feel my wrath, and I'm going to enjoy watching them suffer. They killed two innocent people because of me, oh Gods, it's my fault!" Lillith felt weak in the knees and fell to the ground.

Andrew knelt down and hugged Lillith. "It is not your fault; they wanted to hurt you, stop you from saving the world. You have to stay strong for the world."

"I will be strong and ruthless when I destroy those evil creatures that used to be human. Power hungry disgusting pieces of crap, no one can stop me, I'm ready to destroy evil!" Lillith growled through tears.

"We all will destroy all evil. I love you Lillith."

"I love you too, my sweet, sweet man." They hugged and then kissed through tears.

****

They had one large casket for Lillith's parents, and they could be put in the ground holding each other. Andrew's dad built a beautiful casket that they would be able to fit in perfectly. Lillith aligned the casket with cushions and silk to cover the cushions. It was a beautiful casket. The funeral was outside, the casket was open, and Lillith's parents were holding onto each other. After a few words, they closed the casket, then nailed it closed. This made Lillith break down. Andrew held her tight. Everyone got up and put a flower on the casket. It was packed with flowers, overflowing with them, because it seemed like their whole country had come to pay respects to them. Lillith put a huge sunflower on the casket, knocking flowers off, which made her feel bad, but everyone understood. They lowered the casket into the hole, using ropes. They all started leaving. Lillith was happy that the grave was in the shade of a big tree, and she bought a large tombstone. So you could see it anywhere when you were in the graveyard.

"Let's go, sweetheart," Andrew said, while everyone was leaving. His arm was around her shoulders.

"Okay..... " Lillith agreed. Dom, Selene, Jack, and Jade were by her side. They all went to the pub. When they sat at the corner table, all the dragons jumped off their partners and landed on Lillith, purring and kissing and cuddling Lillith.

Lillith smiled and cuddled all the dragons. The group smiled. "They all love you, Lillith, just like we all love you," Seline said.

*It feels great to be cuddled by tiny dragons." She said while petting them all. They all had a couple of alcoholic drinks, because it had been so long, and they wouldn't be fighting tomorrow. The plan was to relax and just give Lillith all the support she needed.

"I can't eat right now," Lillith admitted to her friends.

"Understandable," Dom said.

Andrew looked at Lillith. "You can have a day without eating, but the next day you are eating, you need your strength."

"I will eat, I will, I just don't have an appetite right now."

"I know, but we will get you as strong as possible," Andrew replied.

"We have to kill the last camp, the day after tomorrow, and as I said, we all attack the skinwalker together. We will overwhelm it, and I know this one is going to be the strongest out of the camps. I want them to suffer. I'm going to find out which one killed my parents. Would the head skinwalker come here? Or was it the other skinwalker? I will get it out of the next skinwalker."

"We will get an answer for you, Lillith," Jack told Lillith.

"Thank you." Lillith tried to smile.

"Can I talk about something I've noticed? Is it okay?" Asked Dom.

"Of course you can, we can't just go on and on about my parents. I love them, but we have to save the world so that I can grieve and plan at the same time, so no worries, guys, I'm stronger than you think." Lillith admitted.

“Well, isn’t it strange that there are five of us, well, supposed to be five, you are the bonus help, Andrew. But each Skinwalker had an element to their Power, fire, earth, air, water, and I think this second last one will probably have powers like Lillith’s.” Dom pointed out.

“Oh wow, that would make sense, maybe they had to find the elements to make the dark crystal? Or, they were given the powers by the leader, the skinwalker. Because they knew we would come after them. Too bad they underestimated us.” Jack agreed.

Lillith looked at their partners. “You guys will have to keep watch on the wendigos, keep them away from us. They are shadows, I don’t know how we are going to fight them, but like before, if the skinwalker is killed, their strange powers go away, and then they are easy to kill.”

The dragons all snorted, agreeing to their job.

Lillith yawned; she hadn’t slept well the night before, as she couldn’t stop tossing and turning throughout the night. Poor Andrew had to sleep on the floor, because her movements kept him from falling asleep, and he didn’t want to get in Lillith’s way. Both Praisia and Bob slept on pillows on the bed, close to Lillith.

“Maybe you should get some rest?” Andrew asked Lillith.

“What time is it, Jack?” Lillith asked.

“It’s six twenty-five,” Jack answered.

“It is too late to sleep. I won’t sleep tonight if I go to take a nap now.” Lillith pointed out.

Lillith thought about it. "You guys, I want to fight the skinwalker. Why wait a day? I'm sad, but I'm also so angry, I want to take that anger out on it. Can we please go fight tomorrow?" Lillith begged.

Andrew looked at Lillith, shocked. "Are you sure you are ready for that?" Andrew was concerned for Lillith.

"Yeah, I have a ton of hatred I want to get rid of. I can use it against the skinwalkers! I feel powerful!"

"Your dad would be proud of you, you never give up, you always fight till the end. You are one heck of a strong person!" Andrew complimented.

Lillith smiled at Andrew. "I learned from my dad. He taught me everything I know about fighting." She laughed, then covered her eyes, and a few tears appeared in them, because she would never learn anything else from her father, nor would she receive any more hugs from either of them. Lillith took a deep breath.

"If you feel up to it, we can go. We are all ready for a fight." Dom agreed.

Lillith smiled at Dom and then everyone at the table. "I'm grateful to have you guys as teammates and friends. I was iffy with you at first, Jack." But I couldn't picture better teammates. They better be prepared, because I'm not normal, you guys should not go easy on them. They are pure evil! We must destroy evil. Just think, after our next fight, we are going to combine our powers to create the light crystal. I'm excited, but a little scared, but we can do it!" Lillith said calmly.

"Yes, we will," all the guys said at the same time.

"Wow, we are all on the same page!" Laughed Lillith.

They talked for a few more hours about nothing important.

****

After they ate, drank, and enjoyed their time taking it easy.

"What time is it, Jack?" Selene asked.

"It's ten twenty-six." Answered Jack.

"We should all go to bed." Andrew proposed.

"Immediately, yes!" Said Jack.

The dragons left Lillith and climbed onto their partners.

"Good night, everyone, sleep well," Lillith commented.

They all headed for the inn. Lillith pulled Andrew into the room.

"Whoa!" Andrew was shocked.

"What do you expect? You are the most handsome, kindest, best guy ever!" Lillith laughed.

"I can't wait until we get married, then I can really show you how much I love you. Andrew hugged Lillith.

"Can't wait, I'm glad we made a pact that nothing sexual, until after we are married. Mom loved that about you."

"Making love isn't as big a deal as people say." Lillith smiled, You know, we don't have to wait." Lillith said in her sexy voice.

"I made a promise to your dad and mom, and I will never break that promise. Even though I want to so badly, some things are worth waiting for." Andrew winked at Lillith.

"You are such a good man." Lillith started playing with Andrew's long blond braid. "That is one of the biggest things I love about you, honourable and patient. But we can still cuddle." She jumped into bed after changing into her pyjamas. Lillith patted the bed for Andrew to join.

Praisia and Bob went to the comfy chair and cuddled together.

Andrew got into bed, in shorts and no shirt.

"Are you teasing me?" Lillith asked.

"No, it is just more comfy for me. Should I sleep on the floor again?" Andrew answered.

"No, it's fine, most guys do that anyway, I know my dad did. I just wanna cuddle." Lillith admitted.

Andrew climbed into bed and lay on his back. Lillith moved over and snuggled with her future husband.

****

The next morning, Lillith was pumped and ready to take her anger out on the skinwalker. Andrew took a shower after Lillith.

“Okay, let’s go to the pub!” Lillith said excitedly.

“Okay, let’s go!” Andrew cheered.

As they entered the Pub, Jack and Dom were waiting in the corner.

“How long have you been waiting for?” Lillith asked.

“Maybe half an hour, no biggie. “ Dom said.

“Dom woke me up....I’m tired and cranky!” Pouted Jack

“Jack, you have to be ready to fight! Drink some coffee! Black coffee!”

“Good idea. I hope I bite my tongue and be careful of what I say.” Jack hoped.

“I hope so, too, because we women will destroy you,” Lillith warned.

That put a smile on Jack’s face. “I needed to hear that.”

"Any time, I will straighten you out." Lillith giggled.

Jack stretched, then asked the waitress for a large black coffee.

Selene and Jade entered, stretching and yawning.

"Morning, guys!" They both said in unison.

"It's about time, we have been waiting for hours!" Jack tried to sound serious, but they could tell he was kidding.

"You are so full of it, Mister!" Jade joked and poked his side.

Jack laughed. "Careful, I'm ticklish!"

"Oh, now I know your weakness!" Jade laughed.

They sat down, ordered breakfast, ate, and then went to the bathroom, preparing for the big fight.

"So, guys, attack the skinwalker all at the same time. He can't take us all on at once!" Lillith ordered.

"That is brilliant!" Seline cheered. "But what about the wendigo's? I kinda forgot last night." Selene felt bad.

"Don't get down, we are a team, our partners are going to keep them away from us."

“What a great idea, and great help.” Otto, you are such a good boy!” She snuggled her partner. Otto snuggled back.

“Okay, the plan is to eat, then relax for half an hour, so we can digest our food, then head toward the shadow wendigo’s. Don’t land until I find the skinwalker, then partners, you get between us and the wendigos.”

They all ordered their food and cream sodas, ate slowly, as ordered by Lillith. Then relaxed for half an hour to digest and just mentally be ready for this fight.

“Lillith stood up and stretched. “Ready, guys?”

Yes, we are.” Jade said excitedly.

“So where is the camp?” Lillith asked.

“It is pretty far, might take an hour to get there, but I found it,” Selene said cheerfully.

Lillith looked at Praisia. “Praisia, do you think you guys are strong enough to make it there in one trip?”

Praisia snorted, then nodded.

“Oh, I love you, my little Praisia.” Lillith cuddled her.

****

They flew for a little over an hour and finally reached the camp. Lillith took out the crystal and saw the red dot on the southeastern side of the camp. “Found it, guys!”

They landed on the southeastern side. They all patted their partners, who made a line between the wendigo and them. The strange thing, though, about the wendigos, was that they weren’t shadows; they just had energy flowing around them, dark energy that made them appear like shadows. They began to walk towards the skinwalker who was staring at them.

Then the skinwalker started cackling, “You look rough, Lillith, you still sad I killed your parents, it was a slow and painful death, it was fun to watch, too bad you slept through it all, you could have saved them if you weren’t a lazy human.”

Lillith glared at it. “Why would you go out of your way to kill two innocent people?”

“Because it was fun, and you are killing my family, so I thought it would be a good way to show you my appreciation of my late siblings.”

“You’d better be ready to die! You scum, you sold your soul for such weak powers. We will kill you not only because you are the enemy, but also because you are a sick twisted freak that killed MY MOM AND DAD!!!!!” Lillith started glowing a light purple. She was enraged; this was going to be payback for her mom and dad.

The skinwalker stood up. “Come at me, I dare you, it will be the last thing you do.” It was threatening.

Lillith lifted her sceptre, while the others took out their weapons and pointed them at the skinwalker.

“Oh, is that supposed to scare me?” It cackled again. Then lifted its hand and shot a lightning bolt at them. They easily dodged it, then Lillith whispered to her team to surround it.”

The skinwalker was walking closer, “Everyone, shoot it with your element blasts. While doing that, they formed a circle around the skinwalker. Lillith thought she might have to attack the skinwalker, because it was blocking attacks physically. How? Who knows, six against one? Yet it’s not even having any effect against it. Lillith walked while attacking it with energy. She met up with Andrew, and she asked him if she could borrow his crystal sword. He handed it to her, then shot more energy at the skinwalker. She snuck up behind it. Just then, the skinwalker turned and grabbed her throat and lifted her off the ground. She was struggling to breathe. She took the sword and chopped off its arm; it screamed in pain. Lillith knew Andrew’s sword was special. Lillith fell to the ground, coughing, still holding the sword.

The skinwalker growled. “It is going to take more than that to beat me!”

Lillith jumped to her feet, and she lost her advantage. “Everyone, use your weapons to attack, not just energy blasts, they have no effect on it!”

“Even if you beat me, you will never win. Our leader has the dark crystal, and he will kill you all.”

“Says the one bleeding out because it lost an arm,” Jack yelled at it.

“Oh, this? It put its hand over the wound and used its energy to stop the wound from bleeding.

Lillith went in to stab it, but it turned so fast, and kicked her away, she crashed into a tree very hard. Everyone screamed. Lillith told them she was fine and hadn’t even been hurt.

Lillith jumped up and yelled, “Attack it all at once, with your weapons. All at once, it can’t beat us all, and if you get knocked down, get up and keep attacking!”

They all attacked it; they were able to chop off its legs and last arm.

“Now what? You gonna bounce and attack us?” Dom said triumphantly.

Lillith walked up to it, then kicked it down so it was lying, and she put her foot over its throat. “I think I will just kill you by choking you. A nice and slow, excruciating way to die. They all looked at Lillith and let her. This was her revenge.

The skinwalker made choking noises.

“What are you trying to say? You can’t breathe? Aww, that is too bad for you, isn’t it?” She stepped even harder, it took ten minutes for it to stop squirming and making choking noises, then it stopped squirming. Lillith took her foot off. The skinwalker coughed, then started laughing. You don’t know what you are against. I’m nothing compared to the leader of us.” It said as it choked.

No, you guys messed with the wrong people, and now you are going to die!" Lillith took the sword and chopped off its head. It screamed, "You are all going to die!" It was dead.

Lillith called over Praisia and ordered her to turn it to ashes." She did as she was told and blew her lavender fire onto the corpse; in a matter of minutes, all of it was ashes. Everyone cheered. Andrew ran to Lillith, hugged her, and kissed her. He was so proud of her.

"You guys better be cheering for all the work you guys did!" Lillith laughed.

As with all the others, the energy swirling around the wendigos was back to normal, and they chopped their heads off. It took them a good hour to get all of them, but they did. All were exhausted.

"Guys, it's just about supper time, let's go celebrate, but no alcohol, we are probably going to be very busy preparing to make the light crystal tomorrow.

****

They all got home safely and were tired. Then they walked to the pub, as their partners shrank and flew to their shoulders, resting.

"I wonder, do you think we can make the light crystal?" Dom asked nervously.

"I believe we can do anything we put our minds to. We are that good." Lillith said.

They all ordered nachos and cheese, slowly eating them, and they were dead tired. They all finished and decided to go to bed; it was getting dark out anyway.

They all went to their rooms and flopped on the bed. "Chopping up wendigo's is hard work!" complained Andrew.

"We did it, though. The head skinwalker is going to be so hard. I need all of your help. I think that was the last of the wendigo's, though."

"Thank the Gods for that," exclaimed Andrew.

"Yea, they were not as violent, I think it's because they were powered by the skinwalkers that ran the camp. Once they died, they lost some of their energy, because they are the walking dead, practically." Lillith pointed out.

"Yea, that makes sense, but I'm excited to make the light crystal, it will make you so much stronger, and with us helping, we can defeat evil and save the world!"

"Hun, do you remember what you agreed to after we beat evil and save the world?" Andrew asked, hoping she remembered.

"Oh, you bet your sweet butt I do!" She giggled. "We are going to get married and live happily ever after. I will take over my dad's gym. What will you do?"

Andrew smiled, "I've been learning to become a carpenter, you know that." Andrew smiled.

“You know my mom already made the wedding dress, it’s beautiful, and fits me perfectly. I’m sad my dad can’t give me away, and that my mom won’t see me walk towards you at the wedding....” Lillith pouted, wiping some tears out of her eyes. “They will be there in spirit.”

“Oh, they wouldn’t miss that, not even death can stop them from going to our wedding.” Andrew put his arm around Lillith and squeezed her shoulder. Andrew looked at his arm, and there was something wet on it. He looked at Lillith’s back, and it was bloody.

“Oh, God Lillith, your back is bleeding. We are going to the Elder witch right now! I don’t want to hear you whine, that is a lot of blood!”

Lillith was shocked; she had no clue she was bleeding. “Okay, let’s go.”

They both walked to the hut and knocked on the door.

The Elder witch opened the door with a big smile. “No more wendigos! Wait, what is wrong?”

Andrew turned Lillith around, and she saw the blood.

“Oh my poor dear, let’s fix that up.” She welcomed them into the hut and sat Lillith down. “Okay, turn around and don’t look.” The Elder witch ordered Andrew. “Unless he has seen you naked before?” The Elder witch asked Lillith.

“Oh no no, we are waiting until after marriage, because we want to strengthen our bond, you don’t need nudity or sex if you can cuddle and chat with. We just decided that, I just believe if we did it early, we would lose interest in the friendship."

"That is beautiful, okay, so turn around, young man." Ordered the Elder witch.

Andrew followed the order. The Elder witch took off Lillith's shirt and bra to reveal a long, deep cut across her back. "Oh dear, you need stitches!"

The Elder witch got out the salve, put it in the wound, and then started sewing up the wound. then added some salve and wrapped Lillith's chest and back.

"This will feel better soon, my dear. Lillith put her bra and shirt on.

"Thank you so much!" Lillith hugged the Elder witch. Andrew turned around.

They both left the hut, said their goodbyes, and went to bed cuddling each other.

Chapter 12 – The Second Search

CHAPTER 12

THE SECOND SEARCH

Lillith woke in Andrew's arms, which was the best feeling ever. Praisa licked Lillith's cheek, and Lillith giggled.

"Oh, Praisia, what would be a morning without you?" She cuddled the little dragon, and that woke up Andrew.

"Good morning, my love," Andrew said softly.

"Are you ready? To join the crystals to make the light crystal?"

"I'm excited and nervous, but anything to destroy the last skinwalker!"

"We are going to save the world! I think if we all work together, we can do anything." Andrew smiled at Lillith. "How is the burn and your back feeling?"

"The burn only tingles a bit, but it's very good, my back didn't even feel like I was hurt."

"You are stronger than anyone I have ever met! Andrew said.

"Well, that is a good thing. I can protect all of you. I'm going to jump in the shower to start my day off on a good note, oh, and this!" She leaned over and gave Andrew a long, passionate kiss.

"Okay, I'm up!" Andrew announced, sitting straight up, and smiled at Lillith.

"Just wait, think of how great it will be when it is all over, and we can get married, it will be so exciting!" Lillith giggled and then called Praisia to come shower with her.

"You are a lucky dragon, Praisia." Andrew joked.

Praisia growled at him, then coed at him, to show she was kidding.

Bob jumped on Andrew's chest and excitedly climbed up, then cuddled his head.

After they were ready, they went to the pub and sat in the corner booth.

"Looks like we are the first ones up!" Lillith laughed.

Jack and Dom came into the pub and waved at Lillith and Andrew.

"Good morning!" Lillith said happily.

"Good morning," both Jack and Dom said.

Are you guys nervous?" Asked Lillith.

"I'm excited. Is this the last thing besides killing the head skinwalker?" Asked Dom.

“I have no clue... but it might be!” Lillith said excitedly.

“What will you do after it’s over?” Jack asked.

Andrew was about to answer, but Lillith beat him to it.

“We are going to get married!! And then, move into my house, I’m going to run the job my dad ran, teaching people to fight, and Andrew will become a great carpenter.”

They ate breakfast and drank some delicious orange juice. They fed and gave the dragons food and water. They were happy, purring and happily walking around the table.

They decided it would be best to get this over with; they were all ready to create the crystal, even though they had no idea how to, and no idea if they had enough power to do so.

They all left the pub and walked to the Elder Witch’s hut, knocking on the door. The Elder Witch opened the door and congratulated them on all their hard work, saying she knew they could do it.

Lillith took a deep breath and asked the Elder witch. “How do we join the crystals? What do we do, we are ready to do it!”

“Okay, let’s see, does everyone have all the weapons?” Asked the elder Witch.

“Yup! They all said.

“Okay, let’s go to the field. I will come and help. This is very exciting, I’ve never seen the light crystal.” She said excitedly

They all walked for a good hour, so no one could interrupt or see what they were doing. Okay, everyone, including you, Andrew, get in a circle. Oh, and the dragons, too, if you could grow big for me? It will help. All the dragons jumped off their partners and grew. Okay, dragons, you guys form a circle around your partners, so Lillith in the middle, the humans in a circle around her, and the dragons in a circle around your partners, including you, Bob.” The Elder told the group.

They all got into position. Lillith was nervous, but she knew she could do it.

Okay, now I’m going to surround you all with salt to keep the negative energy out, including the Head Skinwalker, who has been keeping an eye on you. We can’t let you guys ever be weakened while doing this.” The elder poured salt around the bunch. “Whatever you do, don’t leave the circle, don’t break the circle, just stay as still as possible.”

“How long will this take to do?” Asked Dom.

“It shouldn’t take too long, but it does take all your energy, so I suggest you prepare to be tired.” Warned the Elder witch.

Okay, Lillith, put your sceptre in the air, all others, hold your weapons, I want all of you to picture your energy and powers to flow through you into Lillith. Lillith, you must picture the energy filling you up, and picture a crystal forming through you onto your sceptre. This is key, Lillith, you yourself have the power to form this powerful crystal.”

"Got it." Lillith agreed.

Dragons, open your wings to surround the group and hide and protect them. The Elder witch still could watch; she had a perfect view of Lillith.

"You guys ready?" Asked the Elder witch.

"Ready!" They all said loudly, and even the dragons roared quietly.

"Okay, start!" The Elder witch commanded.

She watched as they all started to glow: Andrew with the lilac-colored energy, then Dom with blue. Then Jade was red, Seline was green, Jack was yellow, and Lillith was white energy. The Elder witch watched in awe at what was happening right in front of her. She waited her whole life to see this.

When they all powered up, they sent a beam of light through them to Lillith, which made the white light glow even stronger. Lillith could feel the powers surging through her; she pictured the crystal forming at the top of her sceptre.

The dragons stood their ground and watched. Lillith could feel each element, especially Andrews, because it was pure love. Lilith lifted off the ground, and she didn't even notice; she was concentrating too much to notice that. The powers merged within her, and she forced them through her into her sceptre, which was now glowing a pure white.

They had been doing this for half an hour, and it was making the humans weak, but they kept sending it. Lillith was in a trance; she was too busy picturing the crystal forming when a huge blast of golden light from Lillith's sceptre shone.

The Elder witch began to cry. This was so beautiful and powerful. They were doing a great job, and she was helping them. She wiped her eyes and laughed. "They are really doing it!" She mumbled to herself.

The energies stopped flowing, and one by one, each one fainted. Lillith was glowing, the new crystal was glorious, a beautiful golden color. Lillith lowered to the ground and passed out. The dragons started making nervous sounds. But they did it! They created the light crystal.

"Calm down silly dragons, it is all part of the process, let's get them to my place." Told the Elder witch. Each dragon picked up and carefully placed them on their back.

Praisia looked at the Elder witch. She wanted to know Lillitrh was okay.

"Don't worry she is fine, just exhausted." The Elder witch said, calming Praisia down. "Let's walk to my place. Maybe they will wake up by the time they get to my place. The Elder witch hoped.

****

By the time they reached the Elder Witch's hut, they were all still unconscious. The elder witch placed hay on the ground to make a comfy bed. "Here, place them here, they will feel much more comfy." Told the Elder witch. The dragons lightly placed the 6 of them on the hay, then shrank and lay on their partners to stop them from panicking. The elder witch placed stones of salt around the hay and sprinkled salt to form a complete circle. There, you should not be able to be spied on; you are all invisible to negativity.

After an hour of waiting, they all except Lillith started to wake up. They all stretched.

"Where are we?" Asked Jack.

"You are safe and at my hut. I'm surprised you guys woke up so fast, because what I read and studied took at least 24 hours, but how could they know? It hasn't ever been needed before." The Elder witch pointed out. "You five are stronger than you know. Andrew looked at Lillith, then at the sceptre, which now had a beautiful, clear golden crystal at the end; Lillith's hair was now all golden. "Why hasn't Lillith woken up like we did? Is she okay?" Andrew worried, went to her, and took her hand, kissing it.

"Do you realise she formed the second strongest crystal in existence? She had to use all her energy to form that crystal."

"Oh, okay, well, I will bring her back to the inn." Andrew said, picking up the sceptre." The others got up and laughed and cheered. "We did it! We were successful!" They all cheered except for Andrew, who was worried about Lillith. They all walked to the pub.

"I'm going to go to the inn and watch over Lillith until she wakes up."

"I will bring you some food and drink; you must be starving, because I know I am!" Jade offered.

"That would be wonderful, I'm craving fish and chips to be honest," Andrew replied.

"Okay, it shall be done!" She ran into the pub.

Andrew smiled, grateful for Jade's thoughtfulness. He then walked to the inn and entered their room. He softly placed Lillith on the bed and put the sceptre next to her, just in case. He petted her head, just looking down at Lillith. Andrew heard a knock at the door, and he answered. Jade was standing there huffing as if she had run the whole way.

"Thank you, Jade!" Andrew said happily.

"Is Lillith still out?"

"Yes..."Andrew said, pouting.

"Don't worry, she will be up in no time!" Jade tried to cheer Andrew up.

"I will try not to worry, it's just-I don't want her to be sick...." Andrew replied.

"She is just sleeping, Andrew, no need to worry." She hugged Andrew, then ran back towards the pub.

Andrew waved and closed the door. He ate the fish and chips, then drank the cream soda. He looked at Lillith; she was breathing normally, so he lay next to her with the sceptre in between them. He reached over and held her hand. Bob and Praisia landed on his stomach and cuddled with him.

"Aww, you two are so sweet!" He petted them both. They cooed softly and licked his hand. "You know there's some fish and chips left over, why don't you two go eat? I'm sure you two are hungry. " They looked at each other, then flew to the table.

"Come on, Lillith, you better be okay," Andrew said in a whisper.

Praisia raised her head and flew over to Lillith, licking her face.

"What is it, Praisia?" Andrew asked.

Just then, Lillith opened her eyes. "Whoa, I feel so weak," Lillith announced to anyone who could hear.

Andrew drew in a breath and hugged her, pushing Praisia out of the way softly. "I'm so glad you are awake!" Andrew laughed.

"Were we successful?" Lillith asked him.

"Why don't you look next to you?" He smiled.

Lillith looked and saw a beautiful, clear golden crystal. "Yayyyyy!"

"Your hair changed again, not blond, but golden," Andrew informed her.

"What? Noooo, I was just getting used to the lavender." She pouted.

"It's beautiful, Lillith," Andrew promised her.

Lillitrh sighed. "Does it look tacky?" Lillith asked, trying to calm down.

"No, no, it's quite stunning on you."

Lillith blushed. "Well, as long as you like it."

Andrew leaned over and kissed her deeply, running his hand through her hair. "You are always beautiful, no matter what." Andrew said and kissed her cheek, "What I'd do to you if I hadn't made a promise." He leaned over and started kissing her neck. Lillith giggled and then hugged him.

"if only..." She whispered into his ear, kissing his ear, then rubbed his muscular chest.

"Lillith, I want you so badly, but I would wait forever to be with you." He said softly.

Lillith blushed. "You are teasing me!" She laughed, then kissed him and pulled him down on her. They made out for half an hour at least, then Lillith's stomach growled.

"Oh." Lillith blushed.

"Let's go get something to eat." He nibbled on her ear, then got up.

Lillith was frustrated; she was having fun being romantic with Andrew, and she could tell Andrew was sad it couldn't continue.

****

The two entered the pub and saw their friends playing with their dragons. Seline looked up and then waved at Lillith and Andrew. They joined their friends at the table.

"Love the hair." Jade complimented.

"Yeah, I guess, I don't know why my hair keeps changing. I miss my red hair."

"Are you kidding? I'd kill to have golden hair!" Both girls said in unison.

Lillith shrugged. "How are we supposed to find the head skinwalker?"

"We will have to ask the Elder witch tomorrow," answered Dom.

"I just want to save the world, then we can get on with our lives," Lillith spoke with a little impatience.

"Why do you want to rush?" Asked Seline.

"Because I want to marry Andrew and start our lives."

Andrew smiled at Lillith. "It will be over soon, hopefully."

They talked and played with their dragons, laughing because Dom was telling jokes that were so bad. They were relaxing, but they all felt a buzzing in them from the element transfer they did to create the Light crystal. It was beautiful, and it radiated.

Andrew noticed the bandage was off Lilith's arm. "Hun, what happened to your bandage? Let me see the burn.

Lillith showed Andrew that the burn was gone. Lillirth was shocked, so were all her friends. "Just a second!" Lillith pulled both Seline and Jade into the bathroom.

Lillith came out of the bathroom and told Andrew the stitches and cut on her back were gone as well, as if she had never been hurt.

"That is amazing! I'm glad." Andrew said happily, he didn't like seeing Lillith have injuries.

"Jack, what time is it?" Asked Jade, who couldn't help but yawn.

"It is eleven forty-five," Jack told them.

"I think we all should go to bed. Tomorrow, we will ask the Elder witch how to find the head skinwalker," Lillith suggested. She ate a quick sandwich.

“I’m so excited it is almost over!” Jade said while stretching.

They all left and went to the inn.

Lillith climbed onto the bed after changing. “Hun, just think, it’s almost over!”

Andrew crawled onto Lillith. “I can’t wait, but I will.” Andrew kissed her neck and whispered into Lillith’s ear. “I love you, my dearest.”

She giggled and pulled him close to her. “I wish we could get married today, then we could have some adult fun.”

Wait a second, your birthday is tomorrow! What do you want for a gift?” Andrew asked.

“Besides you? I have everything I need, but again, I WANT you.” She giggled.

“You are such a tease.” Andrew climbed off of Lillith and lay next to her. I will make you something nice when I have time. I’m really getting good at carpentry.”

“Oh, I know you are, if you are anything like your dad,” Lillith said, then cuddled with Andrew. Both Praisia and Bob were sleeping on the chair.

Let’s fall asleep and get ready for tomorrow. It’s funny, I will be 19 tomorrow, your birthday is in a couple of months, you will be 21, it’s like we were just little kids.”

It has been great to be lucky enough to have you as my future wife." Andrew said with a big smile on his face.

"You are the best. I would be dumb not to be with you," Lillith admitted.

Andrew hugged Lillith, "You are everything to me, I am a lucky man, you are so different from other girls, in a good way, you are my best friend. I wouldn't trade you for the world!" He said, feeling nice and relaxed.

"You don't know how great you are. I will make you learn how lucky I am."

They held each other, kissing, then closed their eyes, and a few minutes later, they were both out cold.

****

The next day, they all met in the pub and ate breakfast. Lillith told Andrew not to say it was her birthday; she didn't want to make a big deal of it.

"That was a great breakfast!" They all had some pancakes with bacon.

"Okay, let's head to the Elder witches' hut," Lillith announced.

"Yes, let's go and finish up our mission," Dom said.

They all left the pub happily.

****

Lillith knocked on the hut's door. Moments later, the Elder witch answered happily. "Hi!" She said cheerfully.

"How do we find the head skinwalker?" Lillith asked.

"Oh, that's easy!" She exclaimed, then went inside the hut and came out with a purple crystal sphere. "This tracks the dark crystal; it will light up and glow faster the closer you get to it." She said happily.

Lillith held the crystal ball. It blinked once. "It blinked!" Lillith cheered.

It will blink faster and faster until you find it." The Elder witch pointed out.

"I can't wait to kick its butt!" Dom said happily.

"What?" The Elder witch asked.

"When we find the head skinwalker, we are all going to kill it!" Dom said happily.

"No!" The Elder witch said insistently.

"Huh?" Dom and the others look confused.

"Only Lillith is the one going to fight the head skinwalker; you guys are just going to protect Lillith and give her support. But she is the only person who is to fight the head skinwalker." The Elder witch said outright.

Lillith looked at everyone. "I will kick its butt, and we will be done with our mission!"

"Oh, sweet hearts, your mission isn't over yet. You need to save the world. Bring the dark crystal to me, and then we will join the light and dark crystals to create a whole new crystal. I will explain it when you bring back the dark crystal. We have to purify it and remove the evil. The dark crystal isn't inherently evil; it's normally pure, but when used for evil purposes, it becomes corrupted. We need to transform them into a single, pure eternity crystal. Then we can work from there."

"How much more do we have to do?" Asked Andrew.

"Not sure yet, it all depends on the outcome." The Elder witch said calmly.

"Okay, I will fight and beat the head skinwalker and bring the dark crystal back. We can do this. Guys, don't help me, no matter how bad it gets, promise me!"

Andrew was unhappy with this, but they all agreed.

They all left the hut, their dragons grew, and they climbed onto their dragons. Lillith told Praisia to let her know when they need a break from flying. Praisia snorted in agreement.

“Let’s go!” Lillith ordered, and they all took off. Flying high in the sky, Lillith took out the crystal ball. It blinked twice, which made Lillith happy.

It had been a couple of hours of flying. The crystal was blinking faster; they were getting closer, but how close? Praisia looked at Lillith while flying. Lillith knew this was Praisia telling her she was tired. They were flying back and forth all over the continent. So this was going to tire their partners out.

“Everyone, let’s land!” Lillith yelled. They all flew down and landed in a barren field. The dragons rested; they all got off and sat leaning on their partners, except for Andrew, who leaned against Praisia to relax next to Lillith. Lillith rested her head on Andrew. The crystal was slowly blinking; they were getting there. Lillith started feeling very tired. She fell into a deep sleep.

****

Lillith dreamt she was in a dark space, then she noticed a bright light far away, and she started running towards it. But it wasn’t getting any closer, then she saw her father and mother. They smiled at her.

“Mom, Dad! I’m sorry I didn’t protect you!” Lillith cried

“It was meant to be. We are happy and proud of who you have become.” Her father told her.

“But we are so lost, it is taking forever to find the head skinwalker..... what if we never find it?” Lillith cried.

"You will go north, it is waiting for you, so be prepared." Her dad warned his daughter.

"But Dad, how can I get out of this darkness? The light is too far away." Lillith sobbed.

You will get there, in time, just go north, it's ready to face you, and you are the only one who can beat it, and I know you will." He told Lillith.

Lillith ran to her parents and hugged them. "I love you both so much, and miss you even more."

They hugged her, and her mother ran her hand through her hair. "You have such a journey ahead of you. Don't doubt yourself, you are so strong, and your heart is so pure, you will change the world, we are so proud of you, my love." Her mom told her.

Lillith hugged and cried hard at how much she missed her parents. Just then, they started to disappear. "We have to go now, don't lose hope and believe in yourself no matter what!" Her father told her. They were gone, and all Lillith could do was cry hard. She then looked at the white light. She knew she had to get there! So she ran as fast as she could. It got closer now, closer and closer, until she reached it. Then her eyes fluttered open, tears pouring down her face.

Andrew held her tight. "Lillith, what is wrong?" Andrew asked.

"My parents visited me, and we need to go north. Not west and east. North!"

Everyone surrounding Lillith was worried.

Lillith wiped her eyes. I'm ok, it's just, I miss my parents.

Andrew cuddled her. "Our partners need more rest, so just hold me, Lillith."

Lillith held Andrew tightly. "I have to beat the head skinwalker on my own, honestly, don't interfere, guys, even if it seems like I'm losing."

They all nodded,

"We all believe in you, you can beat this creature, we know you have the strength, you can do it," Jack said confidently.

"I will destroy it and get the dark crystal, then we will save the world together. Couldn't have done it without you guys." They all hugged in one big hug.

"We couldn't have done it without you, Lillith," Jade replied.

"Thanks, guys, I believe in myself, and it is all because of you guys. You gave me your powers and made me stronger than I could have imagined."

It started getting dark out. "Time really flies when you are searching for evil." Lillith jokes.

"Oh no, we have to destroy evil..." Selene said worriedly.

“I think we should camp out here for the night. We will start in the morning.”

“I have some food we can have for breakfast, as you know, I carry food with me for emergencies,” Jack told the group.

“Oh, good, did you bring something to drink?” Asked Jade.

“I brought jars of water just in case,” Jack announced.

“Thank the Gods for your thinking.” Jade sighed.

“I am or was a traveller, I know there will be times when I would sleep outdoors and need food and water.” Jack smiled.

“We will find the head skinwalker tomorrow and destroy it; that cruel thing needs to be out of this world!” Lillith spoke up.

“Yes, that is true, we will find it and squish it like a bug!” Dom said happily.

“How much more do we have to do? I mean, after Lillith beats the crap out of it, we bring the dark crystal back, purify it, and then join them together. Then what? How much more can we take?” Jack commented.

“I don’t know, nor do I care, as long as we can do it, I will never give up!” Lillith said sternly.

“Well, I never said to give up; it’s just been a long journey to even get to this point. I mean, yeah, it's worth it, but so much pressure on us." Jack said.

"Don't worry, Jack, we all have each other's back, no matter how long it takes, as long as we have each other, we will make it!" Jade said softly to Jack.

The four went to their partners and tried to fall asleep.

Lillith leaned on Andrew. "What is your dream, my love?"

"My dream? This is my dream, having you by my side always."

"Do you really want to be a carpenter?" Lillith asked.

"Yes, I love carving out designs in wood, making things with my hands. I will make beautiful furniture that will sell for a good amount of money," replied Andrew.

"We don't have to worry about money, hun. We have more than enough to live off of, but if that is your wish, I accept and love it."

The more work I put into it, the more it's worth, but I will give in on the prices, because I don't want to get people to pay what they can't." He murmured.

That's my future husband's heart, pure as gold." Lillith said, kissing Andrew's cheek.

"Oh, in my dream, my dad said to go north!"

"Oh, good, now rest up, I have your back, my sweet angel," Andrew said softly into her ear, and they both fell asleep in minutes.

Chapter 13 – Shadows on the Horizon

CHAPTER 13

SHADOWS ON THE HORIZON

Lillith woke up, and it was so quiet. Andrew wasn't next to her. Where could he be? She looked around, and what she saw was a shock that she nearly died from realising. All her team were dead, heads chopped off, bodies lying here and there, even their partners. Blood everywhere.

"Andrew? Peaisia? What happened?" She cried out.

"Over here," Andrew called.

"I'm coming!" She ran towards Andrew's voice. When she got to him, he was chopping off Praisia's head with Praisia's pain ringing in her ears. "WHAT ARE YOU DOING?" She cried.

Andrew stopped and looked at her. "You made me do this...." He then sliced his throat open and fell to the ground.

"What, no, don't die....." She fell to her knees and cried, What was going on? Why did they have to die, and why did it have to be Andrew who betrayed her? "This can't be happening!" She collapsed to the ground, feeling so lost. Then a voice came from nowhere.

“This is your destiny.” Then the voice started laughing at her.

“NO, NO, NOOOOOO!”

****

“Lillith, Lillith, wake up!!!!” Andrew yelled. Praisia nudged her with her giant head. Lillith was crying and screaming no, over and over, shaking her head.

Lillith’s eyes opened, and they glowed a dark purple.

“Oh, Lillith, what is going on?” Jade looked down at Lillith, very worried.

“Lillith blinked until the dark purple vanished. Lillith blinked with tears pouring down her face. Then sat up, seeing she was surrounded by her team and their partners, including Praisia and Andrew.

“Oh my Gods, you all were murdered by Andrew, then Andrew killed himself, and before he did, he said I made him do it, then a voice said this was my destiny!” She sobbed.

“Hun, your eyes were glowing a dark purple when you opened your eyes.”

Andrew explained.

Lillith grimaced and clenched her teeth. "That bastard must have implanted that thought, it prob thought that would stop me, good luck with that!" Lillith punched the ground, then hugged Andrew. "It was so scary!" Lillith hugged Andrew and whispered in his ear. "I love you, Andrew, I love you forever."

"Right back at you, my Lillith." They let go, then Lillith was tackled by Jade and Selene.

"You really scared us!" The two girls told Lillith, glad that things were back to normal.

"I just need to take those thoughts and remind myself it wasn't real, it was horrible, blood everywhere." Lillith wiped her eyes and sniffled.

"Take your time. He knows we are coming, obviously because he attacked you to scare you." Jack said, understanding that that would be such a shock to her system.

Lillith climbed up on Praisia and just hugged her. "Oh my praisia, you were dead too. I watched you and Andrew die in front of me. I don't know what I'd do without you guys." She snuggled her partner.

"Lillith, are you hungry?" Asked Dom.

"Actually, I am. What do we have?" Lillith wondered.

"Beef jerky, water, and some cheese and bread." Answered Jack."

“Sounds yummy to me,” Lillith said, feeling better than before. The dragons shrank and ate some of everything, and Lillith enjoyed all of it.

Lillith called Andrew, and he went with her. “I wanna snuggle with you for strength, we can do it over here so we don’t bother everyone.”

Andrew agreed, and they snuggled under a huge tree, out of sight of everyone.

Andrew sat and leaned on the tree. Lillith lay next to him, her head on his lap, and rubbed his leg.

“In my dream, you murdered everyone, and I saw you murder Praisia, then you sliced your throat with your sword....”

“Don’t worry, hun, that won’t happen, it was only able to do that because you were out cold, defenceless. You will kick its ass, and destroy it, then we can take the dark crystal to the Elder witch and go from there.”

Lillith cuddled even closer. “I just want to calm myself, holding you, not for too long, we have to get rid of the head skinwalker.”

“Just think, we are almost done,” Andrew said calmly.

“Please don’t help me in this fight, I need to do it on my own, I have so much energy going through me that I feel I could fly,” Lillith told.

“What if you are struggling?” Andrew asked worriedly.

"If I'm struggling, I will push myself harder, don't interfere, or even Praisia."

"But what if...." Andrew started.

"No what ifs, I will destroy it on my own. I can't risk losing any of you; the guilt would kill me. I want you to inform the others, and when we are there, stay far away, so it can't reach you."

"So we are supposed to stand there and watch you fight?" Andrew said, annoyed.

"Yes, that is exactly what I want. If it gets to the point I need help, I will call out for you guys, deal?"

"Yes....." Andrew doubted she would call them, but at least she might.

****

Lillith stretched and walked back to the camp. Andrew had a talk with everyone and their partners, including Praisia. They all looked at Lillith with sad eyes, wanting to help her, but knowing deep down that this was her battle.

"Are you guys ready to go?" Asked Lillith.

"Yea....." They were not happy with this plan, but they had to listen, or else Lillith would probably not let them come.

****

They all climbed onto their partners.

Lillith shouted, "Heading north!" Then they flew off, all of them in a group. It took a while, but the gem beeped faster and faster until they reached a field surrounded by woods on the coast. Lillith saw a dark purple flame and knew that it was the head skinwalker. Lillith ordered them to land in the forest, and they all landed.

"Okay, you guys stay here, I will go destroy the head skinwalker." She walked out of the woods and, as she started walking towards the enemy, she began glowing like the head skinwalker, but in a golden hue, not the dark purple of the original. She took out her sceptre and was growing stronger and stronger. She stopped twenty feet away from the enemy. She raised her sceptre. "You'd better be prepared to die." Lillith threatened.

"I think you are the one who better be prepared." Laughed the head Skinwalker and started shooting balls of dark energy towards her friends. Lillith growled and formed a protective shield around the forest. The balls fizzled when they touched the shield.

"Nice try, how cowardly to fight my friends, you need to pay attention to me!" She shot a blast of light at the head skinwalker. It knocked him down, and he was surprised by how strong she was. It stood up, raised the dark crystal, and sent a wave of energy towards Lillith. She blocked it with one hand.

"Is that your best?" Lillith asked with a laugh.

That pissed it off, and he took the crystal and shot a stream of energy at her. Lillith shot the same beam at the head skinwalker, and the beams met in the middle. They clashed, creating a brief explosion, but now they were fighting power against power.

Lillith noticed the dark energy was overpowering her, but that just made her work harder; she pushed all the pain this head skinwalker caused her, the death of both her parents. A slow and painful death. That drove her to the point of no return, and just then, a blast of energy flew through the energy and almost destroyed the head skinwalker. She started to feel weak, but she knew she had the strength. She fell to her knees, the head skinwalker laughed and tried to force the beam to go closer to Lillith to beat her.

"This isn't over, I have the power to defeat you!" Lillith screams while getting back to her feet.

The head skinwalker shot some more shots at her friends so Lillith would lose her focus. Lillith released her sceptre from one hand and, without looking, shot off equal amounts of light energy at the attack on her friends. She then caught up and destroyed the dark flashes before they could reach her friends.

"I'm going to kill this world, and everyone you love will suffer a painfully slow death." It yelled at Lillith, cackling its horrible laugh.

"My parents were already murdered, and guess what happened to the one who did it?" Lillith yelled at it.

"I told my brother to do it, and let me tell you, it was a very painful death, and that they suffered every last second of their pathetic lives."

"You bastard!" Lillith screamed, then shot ten blasts of light at the head skinwalker while still blasting it with the beam. It blocked most of the shots, but two hit it, knocking it down to its knees. It wasn't as invincible as it pretended to be.

Lillith put more force behind the beam of light, and it nearly hit the skinwalker's head. But it still had power, though Lillith knew she weakened it.

"What's wrong, losing your strength?" Lillith yelled.

"I'm stronger than you; nothing can beat the dark crystal's negative power!" The head skinwalker yelled and stood up.

"No, I don't think you understand." Lillith stepped forward, while forcing more energy out, she kept walking closer and closer until she was face to face. The head skinwalker almost fell backwards.

"You see, there's something wrong with the dark crystal skinwalker. Your evil corrupts it; the dark crystal isn't meant to be evil, but you weakened it that way."

"I....it's meant to be ruled by evil!" The head skinwalker spat at Lillith.

"Wrong, and now I'm going to destroy you, with the strength of pure light!" She screamed and blasted the light so hard that the beam of purple was eaten up and hit the crystal straight on. An explosion occurred, and colored lights flew everywhere. Then, the light surrounded Lillith, and the bright light consumed the skinwalker. Then everything went black to Lillith.

****

"Lillith!" Andrew yelled, running towards the fight scene. The skinwalker was gone; where it had stood was a scorched mark on the ground with a skeletal hand sticking out of the ashes. Lillith was on the ground with her sceptre next to her, the golden crystal was still glowing, but it was fading.

Lillith was passed out, and Andrew felt her neck; her heart was beating normally.

"She should be fine, can't say the same for the skinwalker," says Jack.

"Then why did she pass out?" Selene raised her voice and touched Lillith's forehead; no fever, still warm to the touch.

Andrew picked Lillith up, and Praisia rubbed her huge head on Lillith's head, trying to wake her up.

"Can someone carry her sceptre?" Asked Andrew.

"I will," said Dom. He bent over and picked the sceptre up. It was heavy.

Praisia made Andrew ride her. He climbed on top and made sure Lillith was secure.

"Wait, has anyone seen the dark crystal?" Asked Jack.

"Where did it go?" Jade asked.

"I have no clue, but we should get home and talk to the Elder witch," Andrew said.

"Yea, there is no reason to stay here." Dom agreed.

They all climbed onto their partner and flew back to their home.

****

It was getting dark out by the time they reached home. Andrew landed in front of the Elder witch's hut, hopped off, then knocked on her door.

The Elder witch opened the door and looked at Lillith. "What happened? Where is the dark crystal? Bring her in, and explain, please." She asked.

"It wasn't there, it vanished after Lillith's last attack," Andrew explained.

"Hmmm, she must have purified it. Did you see colors fly away?"

"Actually, yes, right at the end," Andrew answered.

"Then she purified it and made the elements separate. I'm sorry, but you guys have to find them, but this time it will be easy, they will be crystallised, so you just need to find the crystals," explained the Elder witch.

"Another quest..... yay." Said Jack sarcastically.

"It is much easier than if you had to purify the crystal." The Elder witch explained.

"How do we find the elements for the dark crystal?" Andrew asked.

"Don't worry about that for now, you need to rest up after such a battle! You guys won! Now relax for a few days."

"We didn't even battle, only Lillith did." Admitted Selene.

"It was the only way it could be done; if you were there, you could have gotten hurt and distracted her fight, here bring her in and lay her down." Ordered the Elder witch.

Andrew laid Lillith's limp body on the table carefully.

The Elder witch checked for a fever; none was found. "I think she just used up every last inch of energy during that fight. I say take her home; if she doesn't wake up in two days, bring her back. Okay?"

"Okay," said Andrew.

They all left. Andrew picked Lillith up and went to their group outside of the Elders' house.

"She is just exhausted, well that is what she thinks, but yea, let's get her to bed, then we can go to the pub and talk," Andrew suggested.

"Sounds like a plan," they all agreed.

Andrew went to the inn and laid Lillith to rest on the bed, kissing her softly on the forehead. He made Bob and Praisia stay and watch over her. They both nodded and curled up next to Lillith's head and rested.

****

When Andrew arrived at the pub, his dinner was already waiting for him—a large steak and potato dish, served with generous amounts of butter and sour cream.

"Thanks, guys, I'm starving!" Andrew announced.

"We figured, and Lillith probably too," Jade answered.

"Yes, no doubt," Andrew said with a worried smile.

"Don't worry, Andrew, Lillith will be fine. She is so strong and has been through so much already." Dom pointed out.

"I will always worry about my Lillith." Andrew sighed.

"You are such a sweetheart," Jade replied.

Andrew smiled a faint smile at Jade. "She is my soulmate."

"Yes, she is, you both deserve each other," Jade said with a warm smile on her face.

Andrew appreciated the compliment more than Jade could imagine. He felt so unworthy of his Lillith, but she loves him, so he must be doing something right.

****

Lillith woke up to two dragons cuddling her face. She yawned and stretched. “Hey, you two!” She smiled and petted the two sweet dragons watching over her. “I’m guessing I won and everyone is at the pub?” Both dragons coed in unison. Lillith laughed and kissed them both. She got up. “Coming, guys?” Both dragons flew onto her shoulder. Lillith laughed and left her room, on her way to the pub.

****

Lillith opened the pub door and saw her friends talking, and Andrew being quiet. Why? She walked behind Andrew and hugged him. “Why are you so quiet?” She asked.

“Lillith!” Andrew yelped, turned around, stood up, picked her up, and twirled. “Oh my Lillith!!! I was so worried!”

“It’s okay, you think I’m going to leave you?” She laughed. “I’m not letting you off the hook that easily.” She smiled and kissed Andrew, then hugged him close.

Jack cleared his throat, to be an ass, Jade and Selene slapped him at the same time.

“Hey!” Jack complained to the two.

“Don’t ruin it!” Jade complained.

Andrew blushed. “Sorry, I was just so worried.”

“Don’t apologise,” Selene ordered.

Lillith stuck her tongue out at Jack and sat next to Andrew.

“Well, you all ate, didn’t you?” Lillith asked.

“Yes, the Elder witch said you might be out for a couple of days,” Andrew explained.

“No worries, it must be annoying that I keep passing out, will try not to anymore, now that we are done.” Lillith smiled.

“Ugh, we are not done,” Andrew explained.

“What?” Lillith looked at Andrew, confused.

“You must have purified the dark crystal, and it separated the elements from the dark crystal, so we have to find them all and make the dark crystal again, then we have to combine the two crystals, and that is all I know,” Andrew answered.

“Ugh, guess we are going on another hunt, a repeat of the first,” Lillith complained.

“We just need to find the crystals of the elements. I don’t think we need to absorb anymore, because we already have them,” Andrew replied.

“Well, that is good, just need to find them, it will be easier and less stressful without the evil being around,” Lillith said with a smile.

“That is true.” Andrew realised.

"I have a good feeling, it will be easy," Lillith said confidently. "Almost done!"

"Yeah." They all said they were tired.

Lillith yawned then ate real quick, so they all could go rest.

****

Lillith plopped onto the bed and laughed. "I have a good feeling now that all the evil is gone.

"How long do you think it will take the world to come back alive?" Asked Andrew.

Lillith pouted. "That I don't know, I mean, seventy-five percent is a lot to get better, but maybe the witches have something they can do to replenish the world?"

"I hope so," Andrew replied.

"Me too," Lillith said, then pulled Andrew into a hug. Praisia cooed sadly from the chair.

"You two, get your little butts over here and cuddle too!" Lillith demanded sweetly.

They both flew over Praisia on Lillith's shoulder and Bob on Andrew's shoulder. Praisia kissed Lilith's cheek. They all lay down, and eventually they all passed out.

****

Lillith knocked on the Elder Witches' hut door. The Elder witch came out.

"You did it!" She laughed. "Thank you!"

"Yeah, we all did it." She smiled sweetly at the Elder witch. "How do we find the elements that make the dark crystal?"

"Oh yes! Be right back, my dear." She went back into her hut and heard her digging around. The door flew open. You use this compass, the same as before, except you know they are crystallized, so it will be easier to pinpoint your way. Still, these elements are on Chella, the other continent, so it will be easier, but harder, because it will be a different place, but it is smaller than Ethridge, our continent."

"Oh, wow, okay." Lillith looked uncertain at Andrew and the others.

"Don't be nervous, dear, it is just like here, just smaller."

"Okay. Then let's pack up and get going." Lillith said, thanking the Elder witch for the compass that would help them find the elements of the dark crystal. She was honestly nervous. Would they be kind? If only Father were still around to give her advice.

Andrew patted Lillith on her back while they walked back to the Inn. "Don't worry, it's not that different, they are mostly kind, like us, don't be nervous, we got this!"

"Yeah, we got this!" Jade and Selene cheered.

Lillith smiled and laughed.

They all went right to bed because they knew their journey across the great ocean would be a big deal.

****

Lillith woke up and stretched. Today was the day, she looked at Andrew and smiled, but he was fast asleep. She decided to kiss him awake, so she softly kissed him on the lips. It took a couple of minutes, but he realised Lillith was kissing him and he smiled and kissed back.

"Good morning," Lillith said, as Praisia and Bob flew onto Andrew's chest and coed.

"Good morning, you two!" Andrew laughed, then hugged Lillith. "If I wake up like this every morning, I'd die a happy death at a ripe old age." He laughed, then kissed her again. Lillith laughed and petted the two little dragons.

"Today is the day we go across the Great Ocean...... I'm nervous, to be honest."

"Why?" Andrew asked.

"What if the people ask why we didn't save them?"

"We didn't know, if they act that way, we will avoid cities if you want?"

"Yes, that is what I want, because I can't face them…." A tear rolled down her cheek as she smiled at her future husband.

"Okay, I will tell the others, they will understand." He wiped away the tear and stroked her golden hair.

"Thank you, let's get ready, then we can go to the pub and eat a good breakfast." She smiled at him, then got up Praisia jumped onto Lillith's shoulder. "I'm jumping in the shower now."

****

Lillith and Andrew entered the pub; Jack and Dom were already there. Lillith smiled at the two.

"Hey guys! Good morning!" Dom greeted Lillith and Andrew.

"Hey!" Andrew smiled.

"I packed a bunch of beef jerky and water for the trip," Jack told them.

"Oh, good planning, Jack." Lillith praised Jack.

"Just don't think they will have open pubs or even inns over on the other continent." Jack pointed out

"About that," Andrew said. "We are going to stay away from populated areas, okay? So they can't blame us for not doing this sooner, okay, guys?"

"Oh, never thought of that, good idea." Jack agreed.

Lillith and Andrew joined Dom and Jack. Just then, Selene and Jade walked in, and they all filled them in on how they were going to avoid the people of Chella, because they might not be happy with them. They all agreed and then ordered breakfast.

"What time should we embark on our new journey?" Asked Jack.

"Well, we have to fly to the edge of our continent, Ethridge, then take a break, then start going across the great ocean, which we have no clue how far they are away, so we should give our partners plenty of time to rest. So, a couple of hours of rest for them, then leave and hope we make it across." Lillith had to think of how tired their partners could get and hope for the best.

"Well, we flew from here to the north without rest. I am sure we will make it across, no worries, Lillith." Jack pointed out.

That relaxed Lillith some. Praisia chirped at Lillith. She looked at Praisia and knew Praisia was making it known that they could do it. Lillith smiled and pet Praisia. "All of you, including you guys." Lillith pointed to all her friends. "Are the best!" She laughed a real laugh.

"Yes, we are an amazing team!" Jade said excitedly.

Breakfast came and they all ate, sharing the sausage, bacon, and eggs with their partners, which their partners were happy to eat.

"Okay, we will wait half an hour, then we will wait," Lillith announced, then they ordered six large glasses of cream soda and one large bowl of water for their partners.

Praisia and the other partners went to the bowl and drank up all the water, then climbed back onto their partners.

"Fifteen more minutes and we are off!" Announced Jack.

They all headed for the bathrooms. They had all finished their drinks and were now ready to leave.

They all stood up and stretched. Jack picked up his bags of water and jerky.

"Okay, let's go!" Lillith cheered. They left the pub, the partners jumped off their partners' shoulders and grew huge. Lillith climbed onto Praisia, as all the others did, and flew away to the eastern edge of Etheridge.

****

When they got there, they landed on the black ground. Lillith jumped down and touched the ground, which was black and dead, with no signs of new life. She grabbed some of the ground, and it was soot.

"This is horrible, it's been like this for most of the flight, can you imagine what it is like all over the world?" Lillitrh asked the others, and they all just shook their heads. This was a sad reality. The dragons lay down and rested.

"Well, there's nothing to do but rest." Said Selene, so they all rested, leaning their backs against their partners, except for Andrew, who joined Lillith. Lillith rested her head on his shoulder.

"Are you still nervous?" Andrew asked Lillith.

"A little, all this dead ground is depressing, I don't understand why anyone would want this?"

"Well, they are gone, I am sure the world will bounce back, don't worry, my dear."

"I hope so, because we won't survive if seventy-five percent of the world is like this…."

Bob rested near Andrew. Andrew looked at Bob and felt bad.

"Bob, do you think you could shrink and come cuddle with us, so you are not alone?" Andrew asked him. Bob nodded happily and shrank, then flew to Andrew and slept on Andrew's shoulder, the shoulder that Lillith wasn't using, and cooed happily. This made Lillith smile, and Lillith knew Praisia was happy because she purred as soon as her son joined them.

****

They all stretched.

"Okay, let's get going," Lillith ordered that they had to wake Selene and Dom up; they had passed out.

The dragons all readied themselves as their partners climbed on them, and they were off.

Two hours into the ride, the weather started to get rough, lightning started crashing, and it started to pour rain.

"What can we do? Guys, we could get struck by lightning!" Lillith screamed over the roar of the rain. Praisia roared to get Lillith's attention.

"What is it, Praisia?" Lillith asked while wiping the rain from her face. Praisia nodded to the left. There was an island! "Guys, there is a small island, follow us!" Screamed Lillith. They all landed on the tiny mountain of an island.

They all landed on the beach, and it was so loud. Lillith never thought rain could be so loud. There's a cave near the top. Let's fly up there and wait out the storm." Lillith spoke loudly so they could hear her over the rain and thunder. "There's not much room, so one at a time, Selene, you go first, then Dom, then Jade, then Jack, then Andrew, and then I will." Ordered Lilith, so they did that. As soon as they got up there, the partner shrank, and they entered the cave until all were up there.

It wasn't a deep cave, but it covered them from the wind and the rain; they waited out the storm as it raged. But finally it passed in about forty minutes.

"Thank the Gods there was a small island here," Lillith exclaimed.

They all ate some beef jerky and drank some water, then one at a time flew back out to the open, now calming ocean.

CHAPTER 14 – TRIALS OF TRUST

CHAPTER 14

TRIALS OF TRUST

"It's right ahead! We are almost there!" Yelled Lillith.

They all cheered as they knew their partners were getting tired. They landed on the dead ground, and what used to be grass crumbled as they touched it.

"This is so sad..." Selene said, touching the dead earth. As it started to get dark, they set up their camp. The partners stayed big so they could lean on each other, except for Bob, who shrank and flew to Andrew's shoulder. Since Lilith and Andrew shared Praisia, Bob shrank and slept with Andrew on his shoulder. The dragons wrapped their wings like blankets around their partners. It kept them nice and comfy and warm. They all fell asleep; all of them were exhausted.

****

Lillith woke up first. She was determined to find the crystals, but she didn't wake anyone else, as she wanted them to sleep as much as possible before their journey on this unknown continent. Well, it was known, but not to them. She was glad that everyone agreed to stay away from populated parts because she was scared they would be mad, and she felt so bad for them. A whole continent was dead.

Jack was the second to get up; Agreed," Lillith, then led the way to the

Elder's hut.

"Oh, Jack, I had the waitress make us a sandwich for all of us for the first morning: egg, sausage, and bacon. Do you want yours now, or do you want to wait, oh, and milk too?"

"Sounds good, but I will wait for everyone to wake up. I brought some steak for our partners."

"That was so nice of you, Jack! They are going to love it." Lillith said happily.

Everyone else woke up within half an hour. The partners, however, were exhausted and hadn't woken up yet. Everyone wanted their partners to get extra sleep, given how hard they had worked. Normally, they would wake up first, but not today. An hour goes by, and all the partners are up.

"Okay, first we have breakfast, partners, can you shrink? Jack has a surprise for you. They all shrank and flew to Jack, purring and squeaking at Jack. Jack laughed and took out a steak for each partner. They all ate happily, then gave them some water in a bowl, and all the partners drank happily.

"Oh, I brought some breakfast sandwiches for everyone." Lillith laughed and handed each a sandwich wrapped in wax paper. They all devoured the sandwiches and drank the milk. Everyone relaxed for half an hour, and then they were off. Lillith took out the compass and started to look for the crystals. They were flying over a lake with an island in the middle. The compass turned blue, indicating that the water crystal was on that island. Lillith yelled for everyone to land, and they did.

"So how do we find the crystal?" Asked Dom.

"The Elder witch said it would be like me finding the partners, so I'm going to follow my forehead," Lillith explained.

They all climbed off their partners; Lillith already felt warm on her forehead. "Okay, I think I feel it, she started walking to the beach, her forehead got warmer, there was a cave, and she felt it would be in there. Lillith entered the cave with everyone following, Praisia coed and rubbed against Lillith's cheek.

"I'm getting close, my forehead is starting to burn." Lillith used her Light crystal sceptre for light; in the very back, a flicker of blue appeared. "I think I found it!" Lillith ran to it, and there it was — a blue crystal. She picked it up and felt the energy emanating from it. "Dom, here's the water crystal!"

Dom held it, and it made him feel even more powerful. This was great.

"Since it's still morning, we will look for another crystal, but we are going to take an hour break because our partners need to rest. Sounds good?"

Dom smiled at Lillith. "You are the boss, Lillith."

"I wonder what element we will find next?" Jade pondered.

"Who knows?" Jack said.

****

After resting, they climbed onto their partners and continued north. As they were flying, they saw a big mountain,

Lillith looked at the compass, and it shone red.

They flew to the top, where they found another cave. Lillith had a good feeling, so they all landed. "This is a volcano that went to sleep." Jack pointed out.

Lillith's forehead started to get warm again. It is the right place.

"Jade, your element is in the cave!" Lillith said excitedly.

Jade squealed. "This is going by so fast!"

"I'm so happy it is!" Lillith cheered.

The entrance to the cave was small.

"Okay, guys, just Jade and I will go in, for safety purposes, okay?" Lillith said.

Andrew hugged Lillith. "Be careful, my angel," Andrew said, but knew she would.

"I will," then Lillith kissed his cheek.

Jade laughed at Jack's sour face.

Selene told them all to be careful.

"We will," both Lillith and Jade promised.

"Okay, let's go, Jade," Lillith said.

"Coming," Jade said, running up to Lillith as she walked fast towards the cave.

Lillith took a step into the cave. It was dark, so Lillith used the sceptre for light again. It was cramped, but since they were the two smallest in the group, they managed to fit in. Lillith's forehead got warmer and warmer until it started to burn.

"It has to be around here somewhere. Look around, Jade." Ordered Lillith.

"I think I see it!" Cheered jade. "It's buried under the soot." Jade dug where she saw a tiny glow, and she found it. "I found it, Lillith!" Jade said proudly.

"Good! Let's get out of here." Lillith said, smiling.

They exited and showed the round crystal sphere; it matched Doms.

"Look at them, they are swirling inside of the spheres, like they were alive, not a heartbeat, but a flow of magic." Said Dom.

"Well, I would hope so; they are filled with magic," Jack spat at the two.

"Look who's jealous." Laughed Jade, Selene joined in.

"Hardly," said Jack in a snotty way.

"Come on, stop fighting, or joking, let's get along," Andrew begged.

"Yea, come on, guys, let's get along. Everyone is going to have a crystal; we are all equal." Dom said wanting peace.

"I was just poking fun at Jack, I wasn't meaning it," Jade said as she carefully put her crystal into her satchel.

A strike of lightning crossed the cloudy sky, which none of them noticed was getting dark out before it was due.

"Oh Gods, we all have to get into the cave for cover!" Yelled Andrew.

"It's just tight in the opening of the cave, lots of room in the cave for all of us," explained Lillith.

Lightning struck a dead tree near the bottom of the mountain.

Jack jumped. "Okay, let's get into the cave!"

The partners shrank and flew to their partner. They all went in; the largest of them (Dom) had to squeeze in. They all sat down. Lillith sat next to Andrew, who lovingly kissed her forehead. "You have been such a great leader. I'm so proud of you," Andrew whispered to her.

"I couldn't have even started if it wasn't for you. You have been cheering me on from the start, but I'm not surprised. No, you have been by my side since I was five. You are my best friend, my soul mate." She leaned on him and kissed him. The two cuddled each other.

"Ugh, Lillith, it's really dark in here," complained Jack.

"Oh yea." Lillith took out her sceptre and asked it to glow a nice warm light. It began to glow nicely. She stuck it into the ground so it would be in the middle. Another lightning bolt cracked across the sky, hitting something, no doubt.

"It's funny, it never rained back home throughout our journey, now the second storm in two days.

"Well, our storm season isn't till leaves start to fall off and die, maybe they get hit with many storms, maybe it's their rainy season?" Selene pointed out.

"I dunno, maybe it's the planet yelling at us for not saving this continent," Jade commented.

"I never would have thought that, but from what I heard, it had started when we were too young to do anything about it," Lillith replied.

“Oh, well, that’s not our fault.” Selene piped in.

“No, it’s not, but I still want to stay away from places where people are.” Admitted Lillith.

“Okay.” They all agreed.

“Guys, I think we are going to be spending the night here, this storm is really getting loud, and I bet it will be like this all night. Are you guys ok to sleep here?” Lillith asked.

“Like we have a choice.” Jack snarled.

“Jack, don’t be such a jerk.” Yelled Jade.

“Hmph.”

“Jack, are you okay?” Lillith asked, because he had been nicer lately.

“I don’t like being closed in a small cave; they freak me out, I’m sorry.”

“It’s okay, we understand now.” Lillith smiled at Jack. Jack smiled back.

“Is anyone hungry? Jack asked.

Everyone agreed, including the partners.

Jack handed everyone a piece of jerky including the partners. Then passed a bottle of water so everyone could drink, and a bowl of water for the partners.

****

The storm died down, and soon it stopped raining.

“I guess the storm wasn’t as bad as I thought.” Lillith pointed out.

“It’s getting dark out, but I think we should get outta this cave.”

“I think that is a great idea.” Jack agreed.

Everyone agreed, so they left the cave and flew north. Lillith spotted a dead forest and decided to land there. Praisia followed suit and landed at the forest’s edge. All the plants that once were green and beautiful were now blackened and brittle.

“What a shame,” Selene said sadly.

There was a lot of dead wood. They used their magic to cut down a dead tree, and then Jade used her magic to start the fire.

Lillith took out her compass to see if there were any elements around them. The compass didn’t show any color, so there were none in this area.

“Three more to go!” Lillith announced.

“I feel we will find them soon, just have a gut feeling,” Jack said.

They didn’t realise how wet everything was, so how were they going to sleep? They were able to start a fire purely on magic from Jade.

Looks like we are sleeping on our partners tonight." Suggested Jade.

Andrew pouted, and Lillith laughed. "You can live one night without me."

"I will try..." Andrew answered.

"You will be fine, Bob will keep you warm, right Bob?" Asked Lillith.

Bob rubbed against Andrew.

"Yes, I know Bob, you are comfy, and it will be nice." Andrew smiles and hugs his partner's huge head. Bob puffs a blast of air out of his nose in agreement.

It was getting late, so they all decided to go to bed. They climbed on top of each other, and it was comfy. The partners wrapped their wings around their partners to keep them warm.

****

They all woke up at dawn, ready to find the elements.

"Three more to go!" Lillith said excitedly as she hopped off of Praisia. They had all their partners to shrink to eat and drink. They all had some beef jerky and water for breakfast.

"Okay, everyone ready to go?" Asked Lillith.

“Yes, let’s get an early start on this. We will cover more area if we do,” said Dom.

“Sounds like a good plan, plus this land creeps me out, too much death,” Selene admitted.

The partners grew, and everyone climbed on their partners and flew above the dead forest. Lillith took out the compass, but there was no change. However, as they were leaving the area, the compass turned green. She looked down, and there was a clearing in the forest.

“Guys, I think we found the earth element!” They all flew down to the opening, which was ten minutes of flying away from where they slept. “Guess we have to be right over it with this thing?” Lillith questioned. “I had a feeling it would be around here, just a gut feeling,” Lillith admitted.

They all landed in the clearing. Lillith walked around, trying to get a sense of where to go. Nothing so far, but as she approached the darker part of the forest, where a cluster of dead trees covered the area, her forehead began to feel warm. “Selene, this way!” Lillith said excitedly. Selene got excited and joined Lillith; all the others stayed with their partners in the clearing.

Selene was thrilled to receive her element, Earth. Lillith followed the warmth on her forehead, which led her to a tree with a hole in it. Her forehead was on fire.

“It must be in that hole,” Lillith announced to Selene.

"Be careful, you never know what could be in a hole in a tree," Selene told cautiously.

"I'm not afraid." Lillith said happily, "I stuck my arm in, and it was a deep hole. But when I put my arm in, all I felt was a cold, round sphere. I grasped it and pulled out a green crystal ball." "Here's your element," Lillith said with a smile on her face. Selene grinned from ear to ear. She held it and felt the power rolling around inside of it.

"Two more to go!" Lillith cheered.

They returned to the clearing, and everyone laughed and was happy. Now, only Jack and she needed to find their elements. This was exciting. Selene carefully put her element into her bag, and they all climbed onto their partners and flew north. They flew for a couple of hours before landing in a desert for a rest. It was now lunchtime, so they all ate and drank, resting for a good hour. Praisia never left Lillith's side, and none of the partners left their sides. They felt something, but no one noticed, too busy being excited. Jack hoped he was next; he yearned to feel the power the crystals bestowed upon them. He wasn't jealous, he was just excited. Lillith had a feeling they would find her and Andrew's last, but that was fine; there was no hurry anymore, no more evil coming after them.

When they relaxed for some reason, Lillith's forehead felt warm. Lillith took out the compass, and it shone a bright yellow. Lillith was in shock; she hadn't checked because she figured it would be up north, thank the Gods for her forehead.

"Jack, your element is here!" Lillith laughed.

“Really? So soon?” He was so excited, you wouldn’t know he was an asshole if you didn’t know him.

“Lillith grabbed Jack’s hand and pulled him towards the dunes. This action made him blush. He secretly had a crush on Lillith, but he knew she wasn’t meant for him. Still, he could enjoy holding her hand.

“Come on, Jack, this way.” As they passed Andrew, Lillith kissed Andrew’s cheek and told him she loved him. Andrew said he loved her back. This made Jack a little sad, but he wouldn’t want to break them up, not that he really could; they were engaged and very happy. Jack rubbed his face to try to hide his feelings, though he was pretty sure Dom had a crush on Lillith, too.

Selene and Jade were both beautiful, but Lillith was different; she was like a breath of fresh air. The way she led them and never gave up, even when things were going wrong.

Jack followed Lillith, trying not to blush. He had to keep his attitude, because if he didn’t have one, he would seem weak. So he followed her and didn’t say a word. Lillith led the way, her forehead getting warmer and warmer until they came to a huge boulder.

“It’s around here, my forehead is burning,” Lillith told.

They looked for holes in the boulder, but there were none.

“Let me try something, back away from the boulder, I think it’s under this boulder,” Jack explained.

"What are you going to do?" Lillith asked.

"Just watch." He grinned at Lillith. He then concentrated, picturing a tornado around the boulder. He then made the boulder lift up and put it down 10 feet away. Lillith was excited to see Jack use his magic; she was impressed. Lillith's forehead began to burn more. "I think you are right," Lillith said. "You are good with magic already. I don't think I could do that, moving a huge boulder."

"Lillith, you could move mountains if you tried. You are the strongest person I know," Jack said, trying not to sound too nice.

"Jack, you are being so nice. Did you bump your head? How many fingers am I holding up?" She joked.

"Haha, very funny, I can be nice, I've just been alone on my journey for so long, it kinda warped me into an asshole." He laughed, Lillith laughed too.

"Jack, you are not an asshole; you can be a jerk, but who isn't a jerk sometimes?" Lillith told Jack.

"I know I am a jerk a lot, but it's not because of anyone but myself. I don't feel bad, but I really should start trying to be nicer." Admitted Jack.

"Well, that would be awesome, because I know you are a great guy, deep, deep, deep inside, hiding." She joked.

"Haha." Jack laughed sarcastically.

Lillith gave him a hug, then gestured to the sand that was under the boulder. "Let's dig!" Lillith said happily. They knelt down, and Lillith could feel it right below her. She informed Jack, and Lillith wanted Jack to find it so he would feel even happier.

Jack smiled warmly at Lillith, then got to his knees and dug. It was really hard because the sand kept falling back into the hole. But eventually, he felt something smooth. "I think I found it!" Jack almost yelled.

"YES!!! GOOD JOB!" Lillith cheered Jack on.

He used both hands to scoop the round yellow crystal. He felt so warm, and he could feel the energy coming from the crystal. He held it up; he was so proud of himself.

"Now we just need to find the last element, mine, Mystic." Lillith was so excited, but where would it be? This might be a hard search and a long one. But they had already searched half the continent in just a couple of days, so she hoped it wouldn't be that hard.

They both went back to the group, and Jack showed off his beautiful yellow swirling crystal.

"One more! Mine left!" Lillith said excitedly. "Praisia, do you need more time to rest?" Lillith asked her partner. Praisia looked at all the other partners as if they were talking, then she nodded her head. "Okay, let's go, maybe today we can go back home to the Elder witch."

"Yes, I would like to leave and get to your home, Lillith. The dead plants and earth are really sickening to me," explained Selene.

"Well, that's understandable. Earth is your element." Said Dom.

"Okay, let's get going!" Lillith laughed, hugged Praisia's head, and then they all climbed onto their partners, and they flew off into the sky.

****

They continued north, with Lillith holding the compass, just in case. A few hours later, needing to rest, they flew down and landed in a dead field. They all relaxed on the ground. The sky was clear, and it was warming up.

Jade jumped to her feet. "Look at that, guys!" They all looked in the direction; a black cloud began to fill the sky, covering it completely, and making it seem even darker.

"What the hell?" Lillith looked at it, got nervous, and then hugged Andrew.

A crack of lightning shot through the sky, and they all jumped.

Then a voice rang out from the clouds, "You think you defeated evil? I was the one who gave my poor comrades the powers. I taught them everything they knew, told them what to do, and you think destroying them will stop the evil? I'm coming for you, Lillith, you and your little group. Once I find you, I will destroy you and the rest of the world. Do you think those skinwalkers could really find and put the dark crystal together? No, I did it, and I will destroy you and this world; you can't win against me. You think just because you won against evil? Well, you haven't seen anything. When I find you, you will all die a slow and painful death. Then I will finish killing this world!"

As soon as the voice stopped talking, the black clouds started to dissipate and were gone.

Lillith looked at Andrew. "What was that? So the skinwalkers got their power from this person? He doesn't know where we are, but we have to hurry and get the crystals ready. So, whatever he is, it's stronger than any of the skinwalkers. In fact, they got their powers from this thing? I need to get our element so we can create the dark crystal, then what?"

Selene put her hand on Lillith's shoulder.

Jade smiled at Lillith. "We can beat him, or it, or whatever the evil is."

"Yes, we can! I just wonder if the Elder witch knows about this mysterious evil." Selene wondered out loud.

They all tried to relax.

"Just one more element and we can leave this place," Andrew said to try to think on the bright side. "We will all fight at once, no more just you, ok Lillith? And that includes our partners; we need to work together, no more solos." Andrew said firmly.

"Okay, guys, we will all do this together." Lillith agreed, but felt she should do it herself. However, she didn't want anyone to get hurt. If they all worked together, though, they could beat anyone she bet. Plus, they would have both crystals, which could be stronger than the mysterious voice. They all looked worried, but since the voice didn't know where they were, they should be safe getting the last element and then heading home to create the dark crystal. Not to mention their partners fighting with them, they would be even stronger.

"Hey, you ok hun?" Andrew asked Lillith because she was a hundred miles away in thought.

"Oh yea, I'm okay, just thinking." Lillith smiled at Andrew, but he knew she was faking it; he could see she was very worried.

Andrew wrapped his arms around Lillith and whispered in her ear. "You know you can show that you are worried or scared. We won't judge you." He then kissed her cheek.

"It's just, now that my mom and dad are gone, and there's nothing we can do to bring all the dead back, how are we saving our world? Most everything is dead or dying, even if we beat this guy, how is that saving the planet? What good were we to this continent? Let this whole continent die, while we live happy lives on the other for eighteen years? I mean, we were too young, but why did this continent have to be destroyed? If only we had been born eighteen years earlier, we could have saved both continents! I find it cruel."

"Hun, it's not like we had a choice in the matter. Don't blame yourself because of what these evil creatures did to our world; it wasn't our job, no one could have stopped them, and they tried. When we grew up, we were meant to do this. You can't change fate, but you can go along with it, and we can destroy this evil bastard!" Andrew said, his voice rising slightly, as he was frustrated with the points Lillith made, and how true they were.

"Listen, guys, we don't have much time to search for the mystic element, we don't have time to cry, we can cry once we are done kicking this new guy's ass! So let's go!" Dom said, which was a shock, because Dom was very relaxed most of the time.

Lillith looked at Dom, tears pouring down her face. "You are right!" She wiped her eyes."Let's finish what we started!" Lillith's smile was real now; she knew they had a mission, and no matter what, she was going to finish it with Andrew and her friends.

Praisia rubbed against Lillith's face, her giant face making Lillith giggle.

"Oh, Praisia, I love you with all my heart." Lillith hugged Praisia's head and petted her. "My dear sweet partner, you are as loyal as Andrew!" Praisia purred, which sounded much deeper because she was usually tiny when she made that noise. Everyone gathered around Lillith and laughed.

"Let's go!" Lillith cheered, then climbed onto Praisia, as all others climbed onto their partners, and flew off. They headed north, reaching the end of the land, but still found no element. They still had half to go. Lillith made them fly low, just above the trees, so they wouldn't be too noticeable.

They flew to the eastern part of the south. As they were going over a farm, the compass shone a lavender light; it was close to an opening in a forest near the farm. It was still glowing lavender. As they flew down, they noticed the opening had a bunch of tall stones in a circle. They landed and walked into this formation that looked hundreds of years old, if not thousands. Lillith and the others jumped off their partners and walked towards the clearing. In the middle of the circle, there was a hole in one of the rocks. Lillith's forehead was burning.

"I think it's in that hole, because my forehead is on fire!" Lillith said excitedly.

Dom looked nervous. "Lillith, something is telling me it is not a good idea to stick your hand in the hole. Let's try a long stick first. I just have a very bad feeling in my gut." Dom told them.

"Okay, Dom, I trust your instincts, so that is exactly what I will do!" Lillith smiled at Dom. Dom tried hard not to blush. He thought to himself, Why did she have to be so cute? He turned away to hide his face.

Lillith found a long stick and put it into the hole; they all heard a loud snap and crunch. Lillith pulled the stick out, and there was a bear trap in the hole. They all looked shocked; that could have been her hand. Dom looked at the trap and pondered. "This would have taken your hand clear off. They all looked around for any trace of beings, but none were found. Lillith dropped the stick and hugged Dom tight.

"You saved my life! There is no way we could have stopped the bleeding, and Dom, you are amazing!" Praisia rubbed her big head against Dom to say thanks. He couldn't help but blush as he felt Lillith's warm body against him; he wished she were his fiancée.

Dom smiled as Lillith let him go. "Poke another stick just in case, I don't want you losing anything," Dom told Lillith. Lillith agreed and inserted another stick, but nothing happened.

“Okay, it’s safe.” Lillith stuck her arm in and pulled out the lavender sphere that whirled with magic in it. It was beautiful.

“We did it!” They all cheered. “But if that was a trap, does that mean the evil guy knows we are here?” They looked around.

“Let’s just go home and see the Elder witch. I don’t want that evil guy to find us.” Jade said.

“Yes, let’s go home, but it will take a few trips, because our partners need to rest in between.” Praisia stepped to Lillith and growled. Lillith looked at Praisia curiously. “What is wrong, Praisia?” Praisia looked into Lillith’s eyes, and then a phrase entered Lillith’s head. “We can fly home in one trip. We need to get back home as soon as possible.” Lillith was shocked.

“Okay, Praisia said they will fly all the way home, with no stops.” Lillith looked worried, but trusted Praisia. “Okay, let’s go to my hometown to see the Elder witch.

They all climbed onto their partners. Lillith put her crystal in her pouch and climbed onto Praisia, as did everyone else, and then they flew home.

Chapter 15 – Before the Storm

CHAPTER 15

BEFORE THE STORM

It took almost an entire day to get home. They barely made it to the field by the Elder witches' hut, but the partners wouldn't stop for a break. Even when they all told their partners to land, they didn't listen. When they finally landed, they shrank before they could get off. None of the partners got hurt when they shrank.

Lillith picked up Praisia, everyone picked up their partner, and cradled them.

"You did a good job," Lillith told all the partners.

They walked up to the Elder witch's hut and knocked on the door. It was late so that she might be sleeping. But to their surprise, the Elder witch opened the door rather fast.

"You found them all? Good job!" She cheered.

"We found all the elements, but something is happening that we don't understand," Lillith said.

"What is it, my dear?" Asked the Elder witch.

“Well, before we got the last element, black clouds started to appear, until the whole sky was dark, Then a man’s voice started talking, like he didn’t know where we were, but he is coming for us and he said he was behind everything, he gave the powers to the Skinwalkers, said he was coming for us basically.” Lillith explained.

“oh no, he must be the true evil of this, hmmm. So now there’s another thing you need to fight, great.” The Elder witch said, sounding very annoyed.

Lillith was shocked that the Elder witch had no clue the Elder witch didn’t know.

“I guess he was what they talked about as the great evil that will destroy the world. I thought they meant the skinwalkers. I’m sorry for not understanding what was happening.” The Elder witch frowned.

“Can we beat him?” Asked Andrew.

“Oh yes, once you create the dark crystal, you then join the two, and you will be stronger than you can imagine! You will have the Eternity Crystal, plus your friends’ energy and your partner’s to fight with. Don’t worry, never lose faith in your powers, and that goes for all of you.” She looked at all of them.

“Okay then, we will make the dark crystal tomorrow, then if we can, you will teach us how to combine the two,” Lillith announced.

Everyone in the group looked at each other, then nodded. The Elder witch agreed too.

“Okay, we will come back tomorrow, okay?” Lillith asked.

“Yes, yes, so proud of you six, good night.” The Elder witch turned and closed her door. They all walked to the inn, got three rooms, then fell into a deep sleep, with their partners in their arms.

****

Lillith woke up to Praisia giving her kisses. Lillith sat up, and Bob flew to her; they both gave her kisses.

“Oh come on, you silly gooses, why are you being so adorable so early?” Lillith laughed quietly, trying not to wake Andrew up. Bob doesn’t seem to care that his partner was sleeping. He jumped onto Andrew’s shoulder and started to kiss him. Andrew woke up startled, but then realised it was just Bob, so he didn’t mind being woken up by his partner.

“Hey, troublemaker.” He said to Bob. Bob squeaked and nuzzled Andrew.

“Good morning, do you feel ready?” Asked Lillith.

“I’m ready to get everything done so that we can get married,” Andrew replied.

“Haha, what is your big hurry?” Lillith poked fun at Andrew.

“You know I’ve loved you since you hit me with that wooden sword.” He laughed.

“You are funny, it’s been a long time coming with us, hmmm?” Lillith laughed

“It’s been too long,” Andrew answered honestly.

“Well, I’m sorry, but we have a lot to do before we get married, dear,” Lillith replied.

“Yes. You are right.” Andrew pouted.

Lillith leaned over and hugged him, and kissed him.”

Andrew perked up and kissed her back.

“I think it’s time we should head to the pub, I bet they are all there waiting for us.” Lillith pointed out.

“Is it that late?” Andrew looked out the window, and the sun was high in the sky. “Wow, it must be lunch.” He pointed out.

“Yea, I think it is, let’s hurry up and go,” Lillith told him.

“Yes, ma’am.” Andrew tied his hair into a braid and got ready almost as fast as Lillith.

“Okay, let’s go!” Lillith said and opened the door.

“Coming!” Andrew followed Lillith out the door.

****

Lillith and Andrew entered the pub, and Andrew was right, they all were waiting for them.

"It's about time," Jack complained.

"Sorry, we woke up late. It's my fault." Andrew explained.

"No excuses!" Jade joked.

The two sat with their friends, and they all had lunch with some cream soda. Then they left for the Elder witch's hut.

****

They knocked on the hut's door, and she came out smiling. "Are you guys ready to create the dark crystal?" She asked excitedly.

"Yea, we are more than ready," replied Selene.

"Okay, let's go to the field and prepare."

They all walked to the field, and then she put everyone in a circle, with Lillith in the middle. Lillith handed the mystic crystal to Andrew, who smiled at her. She winked at him before returning to the center. The others got their crystals out.

Lillith, go back to Andrew. You all need to bond with the crystals to absorb their energies. Lillith held onto her crystal with Andrew, while the others held theirs.

"Now picture your crystal glowing, and feel the energies entering your body. They all closed their eyes and concentrated, like little balls of light. As they did, each crystal began to glow, and then they pictured themselves absorbing the magic within. They all felt the magic entering their bodies; it was a nice, warm feeling. When the warm feeling went away, they opened their eyes and felt even stronger than before. Lillith looked at Andrew, who looked back at her.

"Did you feel that warmth?" She asked, and Andrew nodded with a big smile on his face.

"Okay, now you all have upgraded your power to twice what it was before. Lillith gets in the middle again. Now raise your empty hands to the sky, all of you picture your crystals combining into one crystal, into Lillith's hands," directed the Elder witch. She then poured salt in a circle around the team.

They all pictured the crystals becoming one within Lilliths hands, then one by one a beam left the crystals and flowed to Lillith's hands, as this was happening the crystals in each ones hands started to disappear, and a crystal, the same size as the light crystal formed, it looked different then it did with the skinwalker, it was still dark purple, but was clearer. They all fell to their knees in exhaustion.

Lillith laughed a tired laugh. "We did it!"

"Oh, Lillith, your hair has changed to dark purple," Jade announced.

"At least it's not gold anymore." Lillith laughed.

They were all out of breath but laughed. They sat there laughing for a bit.

"You guys should go sleep in your beds, you have to combine the light and dark crystals tomorrow." Elder witch proposed.

They were all drained of energy, but they slowly got up and started walking towards the inn. Lillith took Andrew's hand, and they walked together, Lillith holding the purified dark crystal and the light crystal on her back, along with her sceptre.

"You want me to carry that dark crystal for you?" Andrew asked.

"No, it is actually light, just like the light crystal.

"Really? Because that sceptre was heavy when we tried to pick it up, we could barely carry it, don't know if it is the sceptre or the crystal, or both, you are so strong!" Dom admitted.

"I don't feel stronger?" Lillith admitted.

"Maybe it is because they are bonded by you? I mean, we all helped, but you created them. Maybe that is why they don't feel so heavy towards you?" Selene wondered.

"I guess that could be possible?" Lillith thought to herself and looked at Jack, wondering if he agreed; he is a witch after all.

"Well, it kinda makes sense," Jack answered.

“Oh well, it might be right or wrong, it doesn’t matter.” Said Lillith, “What matters is tomorrow we get these two crystals to form the eternity crystal, which will be stronger than both, then we can get rid of this true evil, the evil that tried to destroy our whole world.”

They all went to their rooms, took off their gear, flopped onto their beds, and fell right asleep. Their partners coed and fell asleep on their partners.

Everyone fell asleep in the afternoon and slept until lunchtime.

****

By lunchtime they all met in the pub, ate some burgers and fries with cream soda.

“I think we should head towards the Elder witches hut now.” Dom suggested.

“I think that is a good idea.” Jade said. “Get it over with.”

“I wonder how to do it, I hope it isn’t too hard.” Selene hoped.

“Well we won’t find out till we get there.” Said Jack. “So let’s get there and find out.”

“Okay, let’s go.” Agreed Lillith, then led the way out of the Elders hut.

****

When they reached the Elder witches hut, they knocked on the door.

The Elder witch came out with a heavy bag of black rocks.

"What are those?" Asked LIllith.

Jack interrupted and said "They look like obsidian."

"That's exactly what they are." smiled the Elder witch.

"They are used for protection." Jack told them.

"You are pretty smart for a wandering witch." The Elder teased

"What are they for?" Asked Jade.

"Well I'm going to surround you guys as Lillith joins the two crystals."

"Oh?" Lillith questioned.

"Yes this is all up to you now, Lillith, the others will be protecting you." Answered the Elder witch.

"Can you explain what we will be doing?" Asked Dom.

Well first Lillith will be in the middle, then you 5 will be putting up your shields that I taught you, then the partners will be surrounding you guys with their shields of energy, like you guys, then i will surround all of you with these obsidian, to protect you even more from that entity that wants to kill off the world, you must stop him."

The group looked at eachother.

"You have to believe you can, have faith in your power, you can defeat all evil, this world depends on it." Explained the Elder witch. "Oh could someone carry this bag of salt, to keep evil out of the circle?" Asked the Elder witch.

"I will." Offered Dom.

Then they all started for the field, but then they noticed black clouds starting to cover the sky.

"Everyone hide in the bushes! He is trying to find and stop us. "Get close so I can make us not be seen!" Ordered the Elder witch. She then surrounded everyone in a bubble, chanting unknown words that Jack didn't understand, but he was impressed.

The dark clouds covered the whole sky completely. "I will find you,and when I do, you will beg for your lives while I kill everyone you ever cared about, then I will slowly kill you. You can't kill me, I've been around for over two hundred years and as soon as I kill this world, I will make my own kingdom and rule over the survivors and watch them suffer as they slowly starve to death. That is a promise.

While the Elder witch chanted under her breath, Lillith leaned on Andrew softly, her hands were full with the dark crystal. Andrew kissed her cheek to make sure she knew he supported her.

Then the black clouds dissipated and they were safe again.

“He seems strong.” Jack said.

“If he can do that, how can we defeat him?” Asked Selene.

“You six have so much power and strength, and once the crystals are combined, you six will save the world, he may be strong, but so are you, you don’t even realise how strong you guys are, do you?”

The six all looked at eachother, then smiled. “We are going to destroy the evil!” Dom said proudly, and they all cheered. They believed in their power, more importantly, they believed in themselves.

“Okay, let’s continue on the field. We have to set everything up.” Ordered the Elder witch.

They walked until they reached the spot that pleased the Elder witch.

“Okay, Lillith, you go in the centre, now you five make a circle around her.”

Lillith went to the middle, and then everyone surrounded her, forming a semi-big circle. She then told them that the partners would have to grow big and surround them. So, all the partners grew and made an even bigger circle. You all know how to make protective barriers around yourselves, right? They all agreed, and the partners nodded in approval.

"Okay, now just do the same as that, but picture it around all of you in a big dome around everyone. Can you do that?" Asked the Elder witch.

"I know we can, we are strong!" Jade said proudly. The partners rumbled in agreement.

Okay, I'm going to surround you with this obsidian, then after that, make a circle of salt, to keep evil out.

The Elder witch placed the obsidian around the big circle, bringing just the right amount. She then ripped open the bag of salt and made a circle around the whole circle. As she made the salt circle, she muttered things under her breath, likely to strengthen it.

"Okay guys, except Lillith, make the protective barrier around the whole group. They all closed their eyes, including the partners. An iridescent color then surrounded them, like a soap bubble.

"Okay LIllith, lift up both crystals and picture them becoming one, feel the flow through them both, picture them becoming one." Instructed the Elder witch.

Lillith lifted her hands, sceptre in one and dark crystal in the other, she pictured them becoming one. She could feel the energies gathering on her sceptre, then let the energy flow freely to each crystal, but mostly the energy of the dark crystal into the light crystal. She could feel the energy building up, it was very strong, and it just kept getting stronger. She could feel herself getting weak, but that wasn't going to stop her. Lillith continued to picture the two crystals turning into one. Then a bright light shone from the sceptre, and the two crystals vanished.

In their place stood a white, shining, beautiful crystal. Lillith's hair had turned white. The eternity crystal had been made! Lillith started laughing and continued until she fell to her knees, then fell over. Everyone heard her fall, so they rushed to Lillith. They saw the clear crystal that shone with all the colors inside.

Lillith whispered to her friends. "We did it." Then she passed out.

The Elder witch went up to Lillith and checked her temperature, she was fine. "She is fine, guys, just used every last drop of energy."

"Oh Lillith, you are so strong." Andrew picked her and the sceptre up. "Are we done?" He asked the Elder witch.

"Yes! It worked; now, we must find the evil creature that is trying to kill the world. You all did amazing, you should be proud of yourselves, you too partners." The Elder witch said with excitement. You may not know this, but you guys have been doing this for several hours."

"What? It only felt like twenty minutes at most!" Jack said, shocked. They were all shocked, but they felt exhausted. As night fell, they all realised it had taken hours.

"Are any of you hungry?" Asked Andrew. They all nodded their heads, so did the partner. "Okay, I will drop Lillith off at the inn, leave her a note if she wakes up, that we will be at the pub. I feel like a burger" He then kissed Lilliths forehead. "Elder witch, how long do you think she will be out for?"

“Could be a day, could be a week, but knowing her, a few hours.” The Elder witch caressed Lilliths face. “I knew you could do this, since the day i saw you when you were a baby, you are stronger than you know LIllith.”

“Do you need help cleaning up the area?” Asked Jack.

The Elder smiled at Jack. “It would be very nice of you to pick up all the obsidian and put it in this bag. Then you can bring it to my place, are you sure you want to be a wandering witch? You would make a great student, when all this is over with.”

“I would love that!” Jack said surprised.

“I had a feeling you would.” She winked at him and handed him the bag.

“We will meet you in the pub once you are done, ok Jack?” Andrew said

“Okay, it shouldn’t take too long.” Jack smiled and then started to pick up the obsidian.”

All the partners shrank and flew to their partners. Praisia carefully landed on Lillith’s chest. Bob squeaked at her from Andrew’s shoulder, but she ignored him and curled up, staying where she was.

The rest of them walked to the pub, since the inn was close by, Andrew told them not to wait for him. He then carried Lillith to their room, uncovered the bed, Laid her down and covered her with the blanket. Praisia moved to her shoulder and curled up again.

“Praisia, protect Lillith ok?” Andrew asked, Praisia nodded, and then created a shield around Lillith. “Good idea Praisia, thank you.” Andrew smiled, Wrote a note, then left, he was pretty hungry after all that.

Andrew entered the pub and saw they had all waited for him. Jack wasn’t back yet. “We should wait for Jack to come back.”

“Sounds like a good plan, who would have ever guessed Jack would want to help the Elder witch? I think he is getting to be a kinder person,” Jade said.

“I think he is tired of being a travelling witch, I think he has seen enough witches to want to become an Elder himself one day,” Selene explained.

“That would be wonderful! Lillith and I have plans too. Lillith wants to take over what her dad was doing, training and fighting. I want to be the best carpenter, like my dad.” Admitted Andrew.

“I think all of us want to move here, it’s not too crowded, and I want to keep in contact with you guys.” Said Selene

“Do any of you have plans when this is over and done with?”

Jade smiled. “I wonder if Lillith would allow me to train people how to fight with a long sword?” Jade asked.

“She wants to make the building double the size it is today. So I bet she will say yes, and I bet you she will join your classes to learn and master the long sword.” Told Andrew.

"I can teach people how to fight with a scythe, too, for farmers and whoever wants to know," Selene said quietly.

Andrew put his hand on Selene's shoulder. "I bet she would love that idea." I will build a few additions onto the building. Lillith will also probably be taking your lessons, too. Lillith loves to learn physical stuff. You are more than welcome to stay." Andrew said happily.

"What will you do, Dom?" Asked Jade.

Dom thought about it. "I want to stay here, but I don't really have any talents..... Well, I do like making pastries, maybe I will open a shop like that? Does that seem dumb?" Dom asked.

"We don't have any in this town so that you could make a killing. I think I've only had one pastry in my life, and it was a ball - flaky and filled with custard. Tastiest thing I ever ate." Andrew said.

"Oh, that sounds like a cream puff. I have all these recipes from my grandma, who made a living making all sorts of goodies!" Dom smiled shyly.

We will build you a shop, and since everyone else, except Lillith and me, needs a house, it'll be fun to do, at least for me. I love working with wood, creating cozy, comfortable homes.

Jack entered the pub and was happy. "Well, that was fun. Has anyone eaten yet?"

"Jack, do you think the Elder witch will make you her replacement as an Elder witch?"

Jack smiled. “That is exactly what The Elder witch and I talked about; she is sure I will be the best candidate for it.”

“So you are staying in town too? We will all be in one town. We have to stay great friends!” Jade said excitedly.

****

Lillith woke up. It was late, but Andrew wasn’t in bed yet. Praisia coed and rubbed against Lillith’s face. Lillith cuddled Praisia. “You want to go get something to eat at the pub?” She asked her partners. Praisia nodded and coed at her. Lillith got up, put the clear crystal on her sceptre, and then left for the pub.

****

Lillith entered the pub and saw her friends talking and laughing. She smiled and joined them; everyone cheered that Lillith was there.

“Hi guys, I’m glad you are all having fun, but how do we find the evil old guy?” Lillith asked.

“We will ask the Elder witch tomorrow,” Jack answered.

“So what are you all talking about, with such enthusiasm?” Lillith asked

“Oh, just that we all plan on staying in this town, we love this town,” Jade explained.

"I think we are going to have to build the building quite a bit bigger," Andrew said.

"Huh, why?" Lillith asked with a smile on her face.

"Well, Jade wants to teach people how to fight with a long sword, and Selene wants to teach people how to fight with a scythe. So I'm willing to expand on the building." Andrew explained.

"That sounds amazing! We will teach others to protect themselves, I mean, if they ever need it." She smiled at the two girls who looked afraid their ideas would be rejected. But Lillith thought it was a great idea. "I love that idea! But first, you are going to build everyone a house. I will help you, Andrew, and we will have fun."

"I'm so excited!" Jade almost yelled.

"I'm glad everyone is staying in my hometown, and this town is small; there's plenty of room for new houses, and there's tons of room where the building is for two new sections.

"I'm going to be in training to be the next Elder witch!" Jack said happily.

"I know for a fact you will be a great Elder witch! You are so much stronger than you let on." Lillith complimented Jack.

Dom started coughing.

"Are you okay, Dom?" Lillith asked, concerned.

“I’m fine, just cream soda down the wrong pipe.” He cleared his throat. “I’m opening a shop for pastries. I know it seems silly, but that’s really all I’m good at,” Dom explained.

“Dom, I will be your biggest fan and will always eat from your shop; people will come far and wide to eat them, I know it!” Lillith said proudly. “Oh, and you are a great fighter! You learned so fast, be proud of yourself.” Lillith said happily. She was excited to have everyone on their team staying in her hometown. Lillith ate, so did Praisia, and it was getting late.

Andrew stood up. “Guys, we need to get up early, so we should head to bed.” He announced.

“I agree.” Said Lillith, yawning.

“Okay, but you were just sleeping, how can you be tired?” Asked Selene.

“Well, I don’t think I was sleeping; I think it was me letting my body adjust to this new crystal. But guys, be honest, do I look like an old lady because my hair is white? I looked in the window and I was shocked.”

“You look beautiful.” Both Dom and Jack said in sync.

Lillith smiled at the two men.

“Lillith, your hair could be dark grey, and I’d still think you were the most beautiful girl I have ever seen.” Andrew complimented. “You know, hair isn’t everything.”

“Aww, Andrew, all five of you guys are sweet, so grateful you guys are all my friends.” Lillith blushed.

“Okay, let’s go to bed,” Andrew said, wanting to hold his fiancé as they fell asleep. But he was getting a little annoyed, because it had been obvious that Dom and Jack had huge crushes on Lillith. But he trusted Lillith so much that he stopped being annoyed and felt lucky that Lillith loved him.

They all walked to the inn, and as soon as Lillith entered the bedroom, Andrew lay her on the bed, then lay next to her, petting her beautiful white hair, which seemed more like an opal than just white hair. They cuddled as their partners slept on the comfy chair, and they both fell asleep almost instantly.

****

They all woke up at dawn, walked to the pub for breakfast, and then relaxed before heading to see the Elder’s hut. They ordered for themselves and their partners. They left out a bowl of milk for the partners, and they all drank a pint of cream soda before going to the bathroom. They were ready; all that was left was one more thing to do, and then they would save the rest of the world that hadn’t been killed. They left the pub and walked to the Elder witch’s hut, then knocked on the door. She answered with a grin on her face.

“I bet you are wondering where to find this evil guy that manipulates the sky?”

“Yes, we need to know, we are ready for this fight,” Lillith said, ready for anything.

“I’m sorry, I was never taught about this person. I am not sure how to find them. But I have a crystal that detects evil, that could help.” She handed Lillith a blue crystal. “When it turns green, that means evil is close by, so you can search for his whereabouts.” The Elder witch hoped it would help.

“Well, it’s better than nothing. I hope we find him soon. I don’t need this to go on forever,” Jack said, slightly annoyed.

“Thank you,” Lillith said to the Elder witch. “Let’s go fly around and see if he is even on this continent.” They weren’t looking for anything but the last bit of evil in this land, they were ready to fight the last fight. If only there were a real way to talk to this guy. No one, not even the Elder witch, could tell them who and what he was. He was old, over two hundred years old or more, so could he be human? Maybe he was once human and taken over by evil? No one knew, but they had to find him.

Chapter 16 – Destiny Takes Flight

CHAPTER 16

DESTINY TAKES FLIGHT

They flew up into the air. Lillith pointed to the east, as that was the last place the evil was before they destroyed it. Where else could he be? So they flew east, reaching the dead land and landing in a field.

“He is not here, but I have an idea, maybe if we shoot our energy up into the air, he would notice us,” Jade said.

“That’s a good idea, let’s try it.” Lillith agreed with Jade’s idea. “You too,” Lillith told the partners. The partners nodded. So they all aimed up and shot their energy way up into the sky. It was a beautiful sight. Dark clouds billowed towards them and filled the sky.

“I’ve found you.” He roared through the air as the black clouds covered the sky.

“No, we found you!” Lillith yelled.

They all pulled out their weapons, ready to fight. “Show yourself!” Yelled Dom.

“Okay, if you really want me to.” Laughed the evil.

“Were you once human?” Lillith yelled.

The dark clouds flowed down to the field, and a figure started to form. “No, I was never human, but you and I have something in common.” Told the evil one. He finished forming, and he had a six-pointed black star on his forehead. “I’m the same as you are. Except I was left behind with the humans. The black star on my forehead was an omen that I was evil. So humans raised me. I was bad from birth; no one knew how evil I was until I was 15 and killed my foster parents.”

“What do you mean by ‘ left behind?” Lillith asked.

“Well, this world used to be magical, and filled with magical creatures and races, but they left, because of me, but they had no idea that I was the evil they were escaping from.” He started laughing.

“So it took you two hundred years to kill the world, and you didn’t even finish, wow you must be really strong then,” Jack yelled and laughed at the evil. This made the evil one glare at Jack.

“I will kill you first!” He shot a beam of black smoke at Jack, but Jack blocked the attack with his staph. “Is that all you got?” Jack laughed at him.

“How dare you laugh at me!” He yelled, and a large black beam was aimed at Jack. Lillith jumped in front and shot a beam of white light into the smoke, shattering it and striking the evil person in the chest with great force. He fell back, but got up again.

“I will create an army that will kill you all!” He spat at them.

"No, you are not, we are not letting you get away from us. If we have to chase you, we will. This is your end!" Screamed Lillith, then shot him with her beam again, then told all the others to shoot him with their powers. So all the partners and teammates shot him with their beams of the elements. It hit him hard and killed him, but it was different this time. When he landed on the ground, the smoke dissipated, and the star on his forehead turned light blue. He raised his head. "You shouldn't have done that……I was holding it back because I was weak. Now you have to face the real me….."He then turned to ashes and blew away.

"Thank you, now I'm free to kill you all, with my pure evil power." It said, then flew towards them. Lillith and the others put up shields around them, so he couldn't get to them. But he tried to get through the barrier shield.

"This is getting old, you won't live much longer, I'm sucking up all your energies, and it's making me stronger, soon you will be drained, and then I can kill you all slowly and painfully." It laughed.

Lillith was watching all her friends slowly losing the strength, even to stand.

"That's it! I'm getting rid of you forever, no more evil left!" Lillith screamed, she took her sceptre and placed her hands on the crystal, and pictured it glowing, filling it up until it couldn't handle the energy.

Lillith then aimed at the darkness and shot a huge beam of pure white energy at it, and it hit it strongly, but it kept sucking out their magic.

“Guys, keep up the shields, but shoot with all your strength at it, make it come from deep inside of you!”

Praisia roared, and all the partners sent huge beams at the evil, then Jack, Andrew, Dom, Jade, and Selene all did the same.” They shot all their powers at it, and when all the beams hit it, it started to struggle to keep its form. It tried to get away, but Lillith stopped it from escaping.”We won’t let you escape, your time is up!” Lillith cheered.

“You can’t keep this up forever. I will escape! And when I see you again, the tables will be turned!” It bellowed.

“We don’t need forever, you are already dissipating, goodbye evil!” The energy they were shooting combined into a single beam, which then destroyed the beam that was sucking up their energy.

“I can’t move…..” It cried out. It tried to shoot again, but that part of it poofed and was no more. It was shrinking. “No, I can’t die!”

“Yes, you can, you are dying, sorry but not sorry,” Lillith yelled.

It hissed, then, with its last power, shot its last bit of energy at Lillith, but had no effect on her. It started to turn grey, then turned into a white cloud. “I will kill all of you!” It turned into a ball, then exploded into nothing.

Lillith noticed it was gone. “We did it, guys!” She cheered. Lillith turned around and saw that everyone on her team had passed out. Lillith went to Andrew, but couldn’t move, then blacked out.

****

Lillith woke up to find herself on a cot. She tried to turn her head, but the movement hurt. Despite the pain, she continued, and then saw each member of her team lying on cots, all of them passed out. She looked around, recognizing where they were: the Elder witch's hut. But where was Praisia? She couldn't sit up to look, then she felt a little snuggle on her cheek. "Is that you, Praisia?" Lillith asked. Praisia coed. "I can't move, Praisia. Is everyone okay?"

"Oh, you are up, dear!" Said the Elder witch, and walked to Lillith.

"Is everyone okay, and how did we get here?" Lillith questioned, still unable to move.

"Your team, they are okay, just exhausted. Your Partners carried all of you here, thank the Gods they did, I could save you and your team."

"Why can't I move?" Lillith asked, scared.

"Because you used up all your energy on the evil that was trying to kill this world." You have been sleeping for a few days. Here, I will put your sceptre next to you; maybe that will help you? I mean, it's glowing now; there are no such things as coincidences. Oh wow, it's heavy." Complained the Elder witch. Then Lillith felt her sceptre touch her. She felt a warm energy fill her; her pain was gone, and she could move again, slowly. She sat up and cuddled Praisia. "Thank you for helping us, all of you guys, thank you." She spoke to all the partners, and they all flew to Lillith and cuddled with her. It felt good to be cuddled by six dragons.

"If I put my sceptre on each person of my team, will they wake up and be able to move?" Lillith asked the Elder witch.

"Possibly, all this is new to me. I do have something I learned, though, from old journals. You have another quest, and I fear it will take its toll on you, but we can talk about that while everyone else is awake. Try it, dear."

Lillith was shocked. What more can they do? Did they not kill it? Oh well. She slowly stood up, aiming her power towards her teammates. They all began to move, and their partners flew towards them, cuddling against their cheeks.

Andrew sat up and looked at Lillith. "What happened?" He rubbed his eyes. "Last thing I remember was the evil that exploded."

Lillith held his hand. Everyone was now sitting up and looking at Lillith.

"We won, it's gone, our partners brought us to the Elder witches' hut, and now, a couple of days after, I woke and used my crystal to give you guys the energy to wake up."

"Oh, good, we are done with this quest!" Laughed Jade.

"That is not true. There is one last task, but only Lillith can do it, and she needs your support." Interrupted the Elder witch.

Jack yawned while petting Merlo, his partner. "You have to be kidding me, what now?"

“Lillith has to bring the world back to life.”

They all looked shocked. “What?” Lillith was shocked, but if she could do that, it would be amazing!

“You can bring back everything that was lost, including flora and fauna.”

“You are saying I can bring people back to life?” Lillith asked, astonished that she could save the world and bring back the animals and plants that had been killed.

“It says in an old journal that was the true power of the eternity crystal.”

“Wow!” They all said, then stared at Lillith.

Lillith was still looking at the Elder witch. “How do I do that?” She asked.

“I haven’t the slightest idea of how you can do it, but with that crystal and your powers, you will know, just follow your heart, but go relax for a few days, I’m sure you can do it!” Cheered by the Elder witch.

“Wait, that means I can bring back the people who were murdered?” Lillith asked.

“That, I fear, cannot happen. It was their time to go…..I’m sorry.” The Elder witch said sadly.

"Damnit!" Lillith said, almost crying. She really wanted her mom and dad back, but they are gone.

Everyone got up slowly, and all said in unison except for Lillith. "I am starving!"

Lillith looked up. "Yea, me too." She said sadly. "If I can bring back the animals, why can't I bring back the people?" Asked Liliith.

"I don't know, my dear, it just is impossible to bring back the lives of the people on this planet. I wish you could, a lot of good people were murdered, but as I said, it was time for them to pass on. I'm sorry, my dear." The Elder witch explained.

"Okay, everyone, let's go to the pub. We need to talk," Lillith said.

"Thank you for everything, Elder witch!" They all said and left to go to the pub.

As they entered, everyone in the bar cheered at them. How did they know?

A man walked up to Lillith and hugged her tight. She was shocked and embarrassed. Andrew didn't seem too happy with this. She shook the guy off.

"What is going on?" Lillith asked the people. She knew who the man was; she defeated him in a fight and won money because of it.

"You defeated the enemy! You saved what's left of our world!" They all cheered again, and most of them were pretty drunk.

"Yeah, we did," Jack admitted.

They all cheered at Jack.

"Okay, everyone, calm down. We need to discuss things that will help with their mission," Jade said, annoyed at the drunks. All the cheering slowed, then stopped. They went into a booth in the corner to get some privacy.

"Okay, guys, in a few days, I'm going to try to bring back the flora and fauna of this world. I don't know how I'm going to do it, but that's the last thing I have to do; you guys are done. All I need is protection surrounding me, just in case," Lillith told.

"Protection from what?" Asked Selene.

"Well, from bad people, I don't know how long it will take to bring the world back to life," Lillith explained.

Lillith leaned on Andrew, and Andrew hugged her. "I love you, Lillith, now and forever." Andrew kissed Lillith on the forehead. Lillith started crying. Everyone stopped talking and looked at her. "What's wrong?" Andrew asked.

"I just wanted my mom and dad back so badly, I bet they would be proud of me." She spoke quietly so people wouldn't stare at them.

"I did too; they were like second parents to me." He held Lillith and petted her head to soothe her.

Lillith straightened up and wiped her eyes. "Sorry, guys, I'm fine! You guys should be happy, you can drink again!" They all looked at Lillith, not saying a thing. "What?" Lillith asked.

"After you bring the world back to life, that is when we will drink," Dom said, and they all agreed.

"Okay," Lillith agreed.

They all ordered cream soda and their preferred dinner options.

****

Four days later, Lillith felt renewed and ready to bring the world back to life.

"Where do we perform this?" Lillith asked the Elder witch, who wanted to join them and watch Lillith perform her magic.

"You can do it in the field where you guys practised your magic," said the Elder witch.

They all walked to the field. Lillith placed Praisia on Andrew's shoulder and kissed her. They formed a circle, and Lillith stood in the middle, her sceptre raised in the air. Closing her eyes, she focused on her magic. As she let her mind and feelings take over, she turned the sceptre upside down and began to envision the world healing. Picturing the dead starts to heal and turn green again. Then the crystal glowed pure white and flowed energy into the ground. Lillith herself started to glow. Everyone watched as Praisia moved around, clearly wanting to be with Lillith. Still, Andrew gently patted her to calm her down.

They all watched and waited. Time is passing quickly. It had been a few hours of this, just watching the energy flow into the world. Then the glow faded, the energy stopped flowing, and she fell to the ground.

"She used all her energy." Said the Elder witch.

Andrew ran to her and picked her up. She was still warm and breathing fine.

"How will we know it worked?" Asked Jack.

The Elder witch touched the ground and felt the energy of the world. "Selene, touch the ground and tell me what you feel.

"Okay." Selene touched the ground, and she could feel life, growth, and the process of growing. "I feel it, it worked!" The world was healing and growing; dead trees were coming back to life, and animal life was returning. Selene was shocked at the difference Lillith had made.

"Your work as a team is over." Announced the Elder witch.

It was now evening, and everyone went to the pub to celebrate, except Andrew and Lillith. He wanted to be there when Lillith woke up, so he lay in bed cuddling her, making sure she was good. Her hair was slowly turning red again. Andrew brushed a few hairs from her face. Then kissed her softly. Praisia curled up on Lillith's shoulder as Bob cuddled on Andrew's shoulder. Andrew fell asleep; it was getting late. Jade and Selene were already in bed in their room at the inn. Jack and Dom were on their way to the inn, and they refused to drink without Lillith.

****

Andrew woke up to Lillith staring at him lovingly. "Good morning." She said happily

"Lillith, you did it, you made the world come back to life!" Andrew smiled at her. He was so proud of her.

"I figured as much, since I woke up in your arms." She smiled and cuddled him. "Finally, we are done." She squeezed him, and Andrew squeezed back."

"Wait, where's Praisia and Bob?" They asked Lillith, who looked around and saw the two pawing at the window. "What is going on? What are you two doing?" Lillith asked.

Praisia and Bob were now at the door; they wanted to go somewhere. "We should go with them," Lillith suggested. So they opened the door, and the two flew out. They noticed all their friends outside with their partners. "What is going on, guys?" Lillith asked.

"They all wanted out. Is something happening?"

“I have no clue.” Admitted Lillith.

All the partners grew big and lowered themselves for their partners to climb on, so they did. The partners flew off, unaware of their destination, but they didn’t care; they trusted their partners. They flew for an hour, and they all noticed how beautiful the world looked. It was beautiful before, but now it’s even more so. They reached an opening surrounded by forest, and the partners landed and lowered again so they could climb off.

“What is going on, Praisia?” Lillith asked, and Praisia shrank and landed on Lillith’s shoulder; all the other partners did the same. Then, an earthquake occurred, not a major one, but a rumble. Shortly after, a portal of some kind opened up. Lillith and the others were shocked. What was happening?

As they watched the portal now open fully, they saw beautiful birds coming through that were made of fire, all sorts of colors. Then, a few dragons flew through, and they saw people walking through. Some had pointed ears, some were short, some were tall, and some had stars on their foreheads in different colors, but they were similar to Lillith’s.

Lillith looked shocked. “Is this the magical beings that left because of the evil returning?” Lillith asked Praisia. Praisia purred and cuddled her face. While all these beings came through, they saw Lillith, and what seemed like the leader of the bunch walked to Lillith. It was a beautiful woman with pure white hair and a gold star on her forehead.

“Hello, so you were the baby that was dropped off 19 years ago by your mother.” The woman said.

Lillith gulped and nodded her head.

“Your parents, blood parents, want to meet you. They are very proud of you and your team. We are all here because of you; we can return to our natural home. I am the queen of my people, and they are all coming home. We cannot repay you for this honour, but we all give you gratitude for what you six accomplished. Such great tasks made the world pure again, and we could come home.”

“I-I’m glad we could accomplish these tasks.” Lillith smiled.

“My name is Lillian. What are your names?” She asked the group.

Lillith replied. “This is Jack, he rules air. This is Selene, who rules the earth. This is Dom, who rules the water. This is Jade, she rules fire, and this is Andrew, my fiancé. We share the mystic element together, and I’m Lillith.”

“What a great-looking team, you all saved the world and returned it to its natural form. Magic has returned because of you six. We will never be able to pay you back, but we can honor your greatness.” The king joined them. His star was silver.

“Hello heroes of the world, it has been too long since I walked this planet. Thank you for being chosen and for not giving up, even when it seemed impossible, which I’m sure it has felt that way numerous times. We will honour you all greatly.

Lillith looked at the queen, who wore a beautiful golden gossamer dress, and the king, who had a very nice outfit; he looked royally dressed.

Two people started walking towards them. The male had a blue star and silver hair, while the female had a purple star on her forehead. Lillith got an inkling that these two were her parents. Lillith tried to stay calm, Andrew held her hand, and she calmed down.

“Hello, young one.” The male said to Lillith as they approached their group.

“H-Hi,” Lillith said shakily.

The woman, who had the same eye color as Lillith, smiled at Lillith. “You have grown to be so beautiful.”

Lillith smiled, tears appearing in her eyes.

“We are honoured to be your parents,” they both said, bowing.

“I don’t know how I’m supposed to act here.” Wiping her eyes.

“We had no choice, as soon as you were born with the six-pointed star, we had to bring you to this world, you were our future, we never stopped loving you. You remember hearing the voice that told you to go east, that was me, I channelled to you, because I knew in my heart, both of our hearts.” The man held onto his wife’s hand. “Both of us felt you needed help. We could have helped, but that was the only way to communicate, through dreams. I’m glad we could help you.”

“Thank you, it would have taken so long to have found Selene, thank you for the help.” Lillith smiled. “This is my fiancé, Andrew.”-

Andrew smiled and held out his hand to shake Lillith's father's hand. The man took it and shook his hand. "Nice to meet you." The parents smiled at the whole group.

"Oh, how rude. My name is Lilac, your father's name is Sage." Lillith's mom introduced themselves.

"My name is Lillith." Then she introduced her friends and their partners.

"You know your partners aren't regular dragons; regular dragons can't shrink and grow, they were created just for you and your team." Said the queen.

"I figured," Lillith admitted, looking at the portal as more and more creatures and beings came through. The dragons and what she thought were phoenixes flew through. She even saw unicorns coming through. Hundreds of beings were coming through, and just as many, if not more, creatures were following. Then the last of them came through, and the portal closed.

"Well, we have a lot to do, so we will leave you be. We have to find where we want to live, and build our houses." Can we move to where you live, Lillith?" Questioned her mom.

"I would love that, you know where I live, you dropped me off there," Lillith told.

"Yes, we will go there and start to build our house." Said Lilac

The queen looked lovingly at Lillith. "We have to check if our castle still exists." The queen laughed.

"There are castles in a few places. Just kick the people out, they were former kings and queens here, but they became corrupt, and we got rid of their power. So they are free for the taking; they don't belong in them." Explained Lillith.

"We will have to check the castles out then, won't we, my king?" The queen smiled at her husband.

"Yes, my dear." The king laughed. The queen whistled, and a golden dragon flew out of nowhere and landed next to her. We will take our leave, and we will see you again soon." She then climbed onto her golden dragon. The king climbed on, too, and they were off.

"Do you two want to get a ride with us?" Lillith offered her parents.

"That would be great!" Said Lilac and Sage.

"Ok, Mom, you ride with me, Dad, you ride with Andrew.

All the partners glided to the ground, then turned huge, and they all climbed on and took off for home.

****

Everyone in the town, along with people from other towns, came to help them build houses and expand the fighting building with two additional buildings. They tore down Lillith's old house and built a brand new, big house. Everyone had a house within a month's time, on both continents. Farms were back to normal, and there were no more food or drink shortages on the smaller continent. The world was back to normal, and everyone was glad for that. Lillith's blood parents had a place near the Elder witch, which was also near Jack's house, as he was training hard to take over the Elder witch's job.

Dom opened a pastry shop in a shop that had been abandoned for a few years. After cleaning it perfectly, he installed shelves and counters, and it had a huge kitchen to bake his baked goods. He made blueberry treats for Lillith, as blueberries were her favorite treats.

Jade and Selene had plenty of people who wanted to learn self-defense, and Lillith trained many, including women.

Andrew was an amazing carpenter; he could make the furniture fancy or plain, however people wanted it. He was also a big help in building the houses and Lillith's fighting club in the town.

****

Lillith and Andrew had their wedding, mostly just the people who helped with the fight, but the king and queen of her kind came. They told everyone just to come; they didn't need any gifts. Her birth mom and dad walked slowly towards the elder witch, who was marrying them. Lillith was so happy. They did the wedding. At the end of the wedding, Andrew and Lillith shared a kiss, and everyone cheered, starting the party. Lillith and Andrew then rented out the pub, where Lillith ordered a pint of cream soda. Andrew did too. It was a night they didn't want to drink, so they wouldn't fall asleep. They left the pub early to go to their new home. He carried her into their house. They had to watch Praisia and Bob. They heard a loud crash of lightning, but the sky was clear when they were outside. It started to storm, and Lillith swore she heard a cackle. Andrew told her to ignore it until tomorrow.

Lillith and Andrew looked out the window and saw black clouds. Then, they saw a circle - no, it was a portal. Something had come to their world, and they didn't know what to do. The next day, they visited the elder witch, who was working with Jack.

"Did you guys sense anything yesterday?" Lillith asked the elder witch.

"A bad storm came, but that's all I heard and seen," the elder witch told Lillith, then asked why?

"Because I saw a portal open up, but didn't see who came, but both Andrew and I heard a cackle."

"Maybe it was your imagination, because I didn't sense anything.

"Both Andrew and I heard it, but if no one else has seen it, hopefully it was just a strange thunder."

I hope it was nothing.

"Don't worry," said the elder witch. "It's probably nothing."

Lillith hugged both Jack and the elder witch.

"I hope so," Lillith said.

THE

END

www.ingramcontent.com/pod-product-compliance
Lightning Source LLC
LaVergne TN
LVHW060543061126
851857LV00017B/1877
* 9 7 9 8 9 0 5 5 8 0 1 6 1 *